Drones and the Creative Industry

Virginia Santamarina-Campos • Marival Segarra-Oña
Editors

Drones and the Creative Industry

Innovative Strategies for European SMEs

Springer Open

Editors
Virginia Santamarina-Campos
Conserv. & Restoration of Cult. Heritage
Department
Universitat Politècnica de València
Valencia, Spain

Marival Segarra-Oña
Management Department
Universitat Politècnica de València
Valencia, Spain

ISBN 978-3-319-95260-4 (hardcover)
ISBN 978-3-030-07004-5 (softcover)
https://doi.org/10.1007/978-3-319-95261-1

ISBN 978-3-319-95261-1 (eBook)

Library of Congress Control Number: 2018947740

Foreword

Drones, also called unmanned aerial systems (UAS) or remotely piloted aircraft systems (RPAS)—there are slight differences between them—have traditionally been used for military applications. Over the last decade, improvements in electronics miniaturization, control and perception systems and battery technologies have accelerated the growth of civil drone technologies and applications.

Civil drones are here to stay. They provide unprecedented advantages in certain fields such as aerial photography and filming and aerial inspections, where they have become almost irreplaceable, partly because they involve dramatic cost reductions when compared to traditional solutions. Drones are currently employed in hundreds of applications in different domains. And this is only the tip of the iceberg. All the technology roadmaps agree that drones will have a deep impact on society and that this trend will continue over the next few decades. According to Goldman Sachs, the drone market will reach a total market size of $100 billion before 2020. A large percentage of these sales will come from the military sector, yet the sharpest increases will come from the business and civil sectors, which are forecast to grow at yearly rates of over 15%.

Drones equipped with cameras and other sensors are ideal platforms to gather images and other information from inaccessible locations. In fact, most civil drone applications currently centre on aerial photography and filming. Recently, other drone uses, including transporting objects, logistics and precision agriculture, have begun to attract significant attention and are starting to be exploited. Drones are now starting to interact physically with the environment and perform aerial manipulation while flying, thanks to recent research and development work, particularly in European Framework Programme projects, such as FP7 ARCAS and the ongoing H2020 AEROARMS. This is highly relevant when performing tasks such as repairing, installing and replacing items or performing contact inspection tasks with contact sensors, which is of major importance when maintaining infrastructures (i.e. bridges) and industrial plants (i.e. elevated pipes and tanks in oil and gas industries), as it saves on costs and helps to decrease the number of accidents suffered by humans working at a height. In addition, the improvements made in safety and in the autonomy of drones, thanks to advances in perception and control systems among others, will enable drones to become co-workers that can take an active role in productive processes in close conjunction with human workers, though this would require the addition of "soft" materials. These two last applications are very challenging and are still at a low TRL status, yet the results obtained are promising and will almost certainly become a reality in a few years' time.

There is still much to be done in the drone world. Battery lifetimes still critically constrain their applicability. Improvements in safety, security and data privacy are also essential. New advances in sensors, control, navigation, mapping and indoor localization, communications, obstacle detection and avoidance and artificial intelligence, among others, are required to provide drone autonomy.

Drones are very powerful tools. An international regulatory framework, or at least globally homogeneous national regulatory frameworks, will be necessary to facilitate the suitable development of the sector and future applications, whilst also ensuring safety and security as well as the right to privacy and data protection. This framework should also include unmanned traffic management (UTM) and the integration of drones in air traffic management (ATM). The new European Regulation that is soon to be published will represent a significant step in this direction.

The drone sector also needs to consolidate platforms, electronics and software in easy-to-use, off-the-shelf solutions that can be straightforwardly used by non-experts who are interested in taking advantage of drones as "working tools". Aerial photography and filming represent a very large percentage of the current civil drone market and have become essential in cinematography and creative industries. The book you have in your hands and the AiRT H2020 project have contributed to making drone-based solutions an innovative, daily tool at the service of the highly demanding creative industry.

University of Seville, Seville, Spain   José Ramiro Martínez de Dios
Robotics, Vision and Control Group,   Anibal Ollero Baturone
University of Seville, Seville, Spain

Center for Advanced Aerospace
Technologies (CATEC), Seville,
Spain
8 May 2018

Contents

List of Abbreviations

3D	Three Dimensions
ADVMA	Advanced Materials
AiRT	Arts indoor RPAS Technology Transfer
AR	Augmented Reality
B2B	Business to Business
CAE	Culture Action Europe
CCIs	Creative and Cultural Industries
CIP	Competitiveness and Innovation Programme
CIs	Creative Industries
CISCAC	International Confederation of Societies of Authors and Composers
CORDIS	Community Research and Development Information Service
CSA	Coordination and Support Action
DCA	Department of Culture and the Arts (Western Australia)
DCMS	British Department of Culture Media and Sports
DG	Directorate-General
DoA	Description of the Action
EASA	European Agency of Safety Aviation
ECO	European Cluster Observatory
EEA	European Economic Area
EGNOS	European Geostationary Navigation Overlay System
EU	European Union
EUIPO	European Union Intellectual Property Organization
FCS	Framework for Culture Statistics
FP	Framework Programmes
FRIFF	Flying Robot International Festival
GCS	Ground Control System
GDP	Gross Domestic Product
GNSS	Global Navigation Satellite System
GPS	Geographical Positioning System
GUI	Graphical User Interface
H2020	Horizon 2020

I2C	Central microcontroller to slave microcontrollers
ICT	Information and Communication Technologies
IFCS	Intelligent Flight Control System
IIPA	International Intellectual Property Alliance
IMU	Inertial Measurement Unit
IPR	Intellectual Property Rights
IPS	Intelligent Positioning System
ITN	International Training Network
LEG	European Leadership Group
LEIT	Industrial Leadership
MEMS	Microelectromechanical systems
MSCA	Marie Slodowska Curie Actions
NEM	New European Media Initiative platform
NLOS	Non-line-of-sight
OCS	On-board Control System
PAR	Participatory Action Research
R&D	Research and Development
RIA	Research and Innovation Action
RIS3	Research and Innovation Strategies
RPAS	Remotely Piloted Aircraft System
RTK	Real-Time Kinematic
SBAS	Satellite-Based Augmentation System
SLAM	Simultaneous Localization and Mapping
SMEs	Small and Medium Enterprises
SNA	Social Network Analysis
SPI	Serial Peripheral Interface
SWAFS	Science with and for Society
ToCM	Toolkit of Creative Med
TFP	Total Factor Productivity
UAS	Unmanned Aerial System
UAV	Unmanned Aerial Vehicles
UCD	User-Centered Design
UI	User Interface
UNCTAD	United Nations Conference of Trade and Development
UNESCO	United Nations Educational, Scientific and Cultural Organization
UPV	Universitat Politècnica de València
UWB	Ultra-Wideband
VEM	Virtual Environment Mapping
VPS	Vision Positioning System
VR	Virtual Reality
WAAS	Wide Area Augmentation System
WIPO	World Intellectual Property Organization
XR	Extended Reality

Introduction to Drones and Technology Applied to the Creative Industry. AiRT Project: An Overview of the Main Results and Actions

Virginia Santamarina-Campos and Marival Segarra-Oña

Abstract The aim of this book is to disseminate the results and actions deployed within the H2020 European Project AiRT, Technology Transfer of RPAS (Remotely piloted aircraft systems) for the creative industries. This book collects the different approaches of the project, including the definition of the problem and needs identification, the technological aspects and the business model definition. Different experts from the industry as well as from the academic sector participated in this book by discussing the results of this highly innovative project and how it will impact on different stakeholders, from society to the creative industries, considering the economic impact that the drone sector and creative industries represent in Europe, but also how the successful process of transferring knowledge and technology is supported by the European Commission and from European Universities. This is a key factor when analysing the success of this short and very demanding project.

The original version of this chapter was revised. A correction to this chapter is available at https://doi.org/10.1007/978-3-319-95261-1_12.

The project has received funding from the European Union's Horizon 2020 research and innovation programme under grant agreement no 732433. Reference: H2020-ICT-2016-2017. This book reflects the views of the authors and not necessary the position of the Commission.

This publication has received funding from the Program for the promotion of scientific research, technological development and innovation of the Counsel of Education, Research, Culture and Sport, Valencian Region. Reference: AORG/2018/093.

V. Santamarina-Campos (✉)
Conserv. & Restoration of Cult. Heritage Department, Universitat Politècnica de València, Valencia, Spain
e-mail: virsanca@upv.es

M. Segarra-Oña
Management Department, Universitat Politècnica de València, Valencia, Spain
e-mail: maseo@omp.upv.es

1

1 The Content

After providing an overview of the drones' industry history and its future evolution from a technological point of view as a prologue to the book, in this first chapter a contextualisation of the AiRT project is presented. We start by setting out the identification of the problems on which the proposal was based, its objectives, methodology and work plan development, including the main achievements. The detailed explanation of the project structure and the different work packages, its objectives and resources deployment are presented.

Then, the second chapter analyses the economic impact of the creative industry in the European Union, followed by the third chapter, that presents the study of the creative industries' needs and how their market is expanding and the demand for services related to their activities is increasing.

In the fourth chapter, the authors focus on one of the key points for SMEs for the success of tech-transfer and innovation: financing. They do so by exploring the different European instruments of financing and how they could be applied to the creative industries.

In the following chapter, successful cases where technology and creative industries shake hands are disentangled, with a special focus on the filming industry, where technology is commonly applied.

In Chapter "Storyboarding as a means of requirements elicitation and user interface design: an application to the drones' industry", the storyboarding method is introduced, along with its use as a tool for requirements elicitation, and user interface design in the drone industry is explained. The storyboards are universally understandable and provide a common ground for sharing ideas, as well as for discussing and discovering new points of view. Storytelling in this type of technology transfer projects is very innovative and highly recommended.

Chapter "Usability and experience of the creative industries, through heuristic evaluation of flight software for mapping and photogrammetry with drones" includes the analysis of 30 mesh or mosaic flight plan software programmes, in order to carry out a technical analysis related to aspects of usability and design, with the aim of optimising the design of the software. In this study, a group of experts from the consortium and SMEs from the creative industries and aviation sector participated.

In Chapter "How a cutting-edge technology can benefit the creative industries: the positioning system at work", one of the technical novelties that are integrated in final product is presented: the innovative positioning system developed by Pozyx and its application for the indoor aerial and the creative industries.

In Chapter "Indoor drones for the creative industries: distinctive features/opportunities in safety navigation", an analysis on the indoor drones' characteristics and the differences from the existing RPAS is presented, with a special focus on benefits and future opportunities for the aerial industry.

Lastly, and before the conclusions sections, the links of the industry with the public administration are studied and best practices on pilots' training are presented as an example of co-regulation.

This approach to the project, will aim to provide a framework for understanding the main innovations achieved by the AiRT project, which will be presented throughout this publication. In the following paragraphs the objectives of the project are disentangled, and the definition of the final user needs, as well as the final product requirements are presented in the different chapters, with the objective of spreading knowledge and helping universities and the industry with our experience.

2 The Challenge

The goal of AiRT project is to develop an indoor RPAS built especially for the needs of the creative industries. Currently, different types of drones do exist. However, almost all of them are designed for outdoor applications. And, the few which claim to work indoors cannot be used for professional use. With this project, we want to provide the creative industry with an easy-to-use tool that gives professional results. To do so, our final product is a drone that includes the latest technological advances: a cost-efficient, but very precise indoor positioning system and an intelligent flight control system that ensures safety.

Currently the aviation industry is experiencing an unstoppable revolution with the recent and successful introduction of drones into the mass market. The technological advances of these tools have allowed for the reduction of production costs, which have resulted in a price drop; consequently, drone use has expanded to other sectors, such as leisure, transportation and fleeting of goods, ecological surveillance, etc. (De-Miguel-Molina and Segarra-Oña 2017), thereby decentralising its military use and opening new business opportunities to industries such in the creative sector (Santamarina-Campos 2017).

Technically known as *Unmanned Aerial Systems* (UAS) or *Unmanned Aerial Vehicles* (UAV) and more recently *Remotely Piloted Aircraft Systems* (RPAS), these aircrafts, which operate without a human pilot on board, can be controlled either autonomously or remotely by a pilot. In this context, both outdoor photography and filming, previously reserved for large productions, have become essential resources for advertising campaigns, news on TV, filming of events, film productions or the registration of heritage sites.

As pointed out by a European expert in the field of architecture, arquitects are "super happy" with the possibility of, for example, shots in the world of architecture, which cannot be done without drones. He also mentioned that "the pictures are spectacular, the results amazing".[1] On the other hand, as another expert summarised, the use of aerial images for photogrammetry has contributed through a series of significant advantages, such as greater safety for the operator, efficiency in providing millions of points and a reduction in the execution time and costs.[2]

Nevertheless, in the last few years, a series of events occurred that led to enormous changes in the way of understanding the RPAS market; on one hand, a large number of companies offering products and services around the drones in

---

[1]Informer 6-See Annex 1.
[2]Informer 13.

Fig. 1 The drone market environment map 2018. Source: DroneII (2018)

general, and electric multicopters[3] in particular, have emerged, reshaping and enriching the industry (see Fig. 1). On the other hand, the consequent and hasty legislation development that regulates the use of drones at a global level, led to a reduction of the market size due to the difficulty introduced when carrying out aerial work outdoors (De-Miguel-Molina and Segarra-Oña 2017).

This scenario has turned the (initially small) drone market into a very complicated environment, with huge competition and a much reduced market, due to legislative difficulties.

Despite these circumstances, drones are an unstoppable phenomenon, and we are witnessing only the beginning of their great development. Therefore, at this point we must seek alternatives that allow new applications that facilitate the permanence and development in the market of this new sector, by offering products and services around drones.

A new market niche is indoor applications, where legislation does not apply, and the creative industries sector envisions great advantages: the creative space can be increased, a greater freedom of movement to the camera can be offered, the risks for the camera operator are reduced, invasive auxiliaries can be suppressed, and equipment is affordable. In the opinion of an expert in the field of architecture, a main limitation of filming indoors is the time required for setup. He stated that "you have little time, normally 1 or 2 days at the most, and the blue hour is 10 minutes ... then moving a crane from one place to another in a building ...you would need many devices at the same time to be able to shoot". If instead you work with a drone "it is very easy, because you move it inside and even—I don't know—you can programme it".

Despite the new advantages and creative capabilities offered by drones indoors, their use for professional applications is currently not feasible, since they lack an effective positioning system and safe design.

The vision positioning system (VPS) and the motion capture system are the main positioning systems available on the market today. The first is less expensive; however, it is conditioned by its environment, and normally they are used indoors "... to avoid shocks or obstacles rather than to position themselves ..."; the drone "... can know where it is with respect to a wall with some sensor, with respect to the floor with cameras, ... but it does not know perfectly where it is, like to programme a flight path" (extracted from interview—informant 13). The motion capture system, on the other hand, is very precise. But it needs many auxiliary devices and it is expensive, so it can only be used in big productions.

According to one of the informers of the focus group, the current problem is that there is not a safe design for drones that is adapted properly for indoor use. Thus, there is a latent risk of injury for the operator and the general public, and also a potential risk of property damage, due to the constant proximity of the drone to people and property.

---

[3]Drone is any aircraft that flies without a pilot with different types of engines and 2 or more rotors. We focus on electric engines and several rotors (multicopter).

3 The Goal

The goal of AiRT[4] project (AiRT 2017, Technology transfer of Remotely Piloted Aircraft Systems, RPAs, for the creative industry) is to develop an indoor RPAS especially built according to the needs of creative industries (CIs). The focus was on the development of an easy-to-use tool giving professional results, which will help the CIs to provide new services and improve their competitiveness within the European market. In order to achieve this objective, we retrieved information directly from the final users to finally determine that the innovative indoor RPAS should include:

- **A new indoor positioning system**, which has been identified as the main key element by the European experts who participated in the definition of the needs. Thus, we have an integrated Pozyx system, which is based on UWB technology that has centimetre precision in every axis.
- **An intelligent flight control system**, which integrates IPS, autonomous flight control and 3D reconstruction environment. As a result, the drone is user-friendly and safe, while it guarantees stable positioning in order to achieve high-quality results.
- **The integration of active and passive safety measures**, which constitutes one of the fundamental points not only for the safety of the operators, but also for the preservation of spaces with patrimonial value where the security margin must be extreme. Therefore, our passive security system includes a lightweight and safe design. Moreover, our active security measures include UWB technology, flight path software and an efficient safety system for drones.
- **Professional camera control**. The RPAS includes professional camera control with interchangeable lenses, 3-axis control, command during flight and the possibility of planning the shots. These features were highlighted by the experts in the different focus groups that the team organised.[5]
- **Easy operation.** Hence, AiRT can be controlled on different platforms such as laptops, mobiles or tablets and it allows simultaneous user control. Also, the user does not require any prior knowledge of aviation and photography. It allows an intelligent flight simulation after a 3D reconstruction.
- **Cost-effective**. The RPAS is affordable; at a sum of approximately 9000 € it will be cost-effective compared, for example, to renting a scaffold for 1300 € a day.

---

[4]Definition of AIRT, *chiefly Scottish*: compass point.

[5]Informant 20 said: "...since the auxiliary tools generate numerous restrictions, for example, the pole has many limitations, if you could stabilize the drone at a point [x] and then it would take the shots, it would be very important". Therefore, "if it could be located at a point and could make the 360 degrees panoramic shot ...that would be good." "The more setting options you have, the more useful everything will be, because many times you do not know the problems that you have above and if you have to fly down with the drone and then raise up again—that takes time".

Our specific objectives were to analyse the needs of the creative industries, to adapt the indoor positioning system to the drone, to develop a user guide, and to adapt the pre-existing drone (model called "origami") from AeroTools, (one of the project's partners), to integrate and validate its components, to develop a demonstration with the creative industries and to propose an indoor legislation.

The initial idea was to provide the creative industries with an RPAS for interior spaces, so we can enhance their creativity, enabling them to offer new services. The final drone is expected to fulfil the needs for:

- Publicity agencies that are looking for freedom of movement so they can explore new perspectives.
- The movie industry, which is looking to reduce time and costs of auxiliaries and to develop the pre-visualisation of the storyboard, more specifically camera movements and shots.
- The photography industry that is looking for tools that allow them to explore new perspectives.
- TV production sets, which are always looking for originality and continuously and creativity exploring new techniques.
- The performing arts that are looking for dynamic structures, since their representations are itinerant.
- The video game industry that is looking for affordable and simple solutions to create more realistic video games.
- Architecture and heritage organisations that are looking for quick, simpler and less invasive solutions for supervisions, inspections and graphical surveys.
- Therefore, as previously stated, the concept of the AiRT system has been based on

  - a drone for indoor spaces used by the creative industries, which has a robust, compact and lightweight structure with all necessary safety measures.
  - an intelligent flight control system, which can be used by non-aviation experts on different media platforms.

4 Approach and Methodology

With the objective of transferring the AiRT system to the creative industries in an efficient way, the project has been developed in four interconnected phases, which are the identification of needs, the adaptation and optimisation of different technologies, the integration and validation of the system and finally the evaluation and demonstration of the system. The evaluation has been present throughout the four phases in a systematic way (see Fig. 2).

The methodology used in the development of the project has been based on Design Thinking (Both 2009; Martin 2009; Stickdorn et al. 2011; Johansson-Sköldberg et al. 2013), where the participation of the end-users is present throughout the entire process (see Fig. 3).

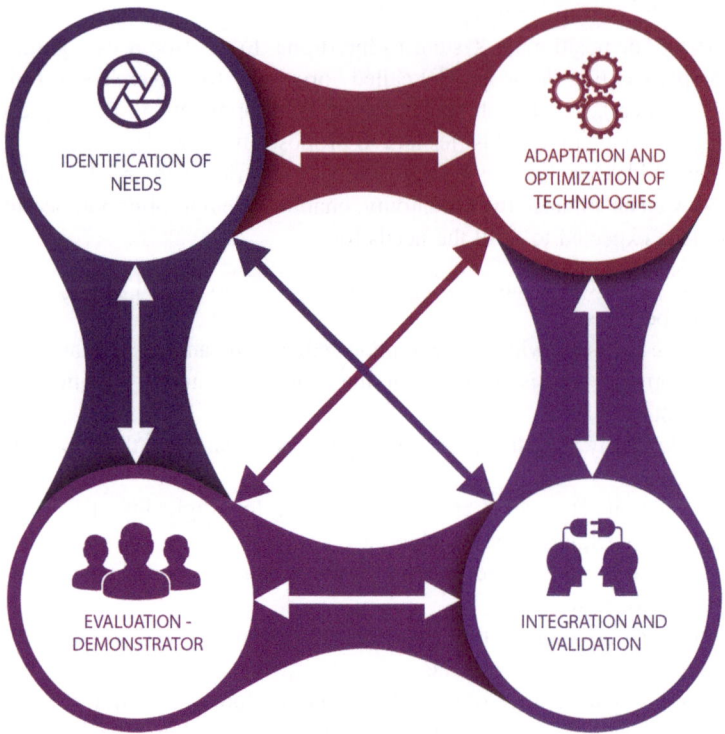

Fig. 2 Approach of AiRT in four phases. Source: Own elaboration

The use of this methodology has been aimed at understanding and providing a solution to the real needs of the creative industries, in order to align the detected needs with what is technologically feasible and commercially viable, thereby increasing their competitiveness (Brown 2009) and defining a valid business model (Osterwalder and Pigneur 2010). For this reason, the process focused on three pillars (see Fig. 4):

1. Empathise with creative industries to understand their problems and requirements.
2. Generate collaborative activities with end-users, in order to obtain creative and analytical feedback, thus creating a systematic process incorporating points of view from all stakeholders.
3. Develop prototypes that will be validated by end-users in order to identify any possible improvements before producing the final product.

Thus, **five interrelated phases** were developed. Interaction with end-users was always present (see Fig. 5).

We followed a participatory process of five steps, where we conducted the focus groups (Step 1). With the content analysis of the information we created some needs maps (Step 2) and then we drew the user's storyboards (Step 3). Next, we designed

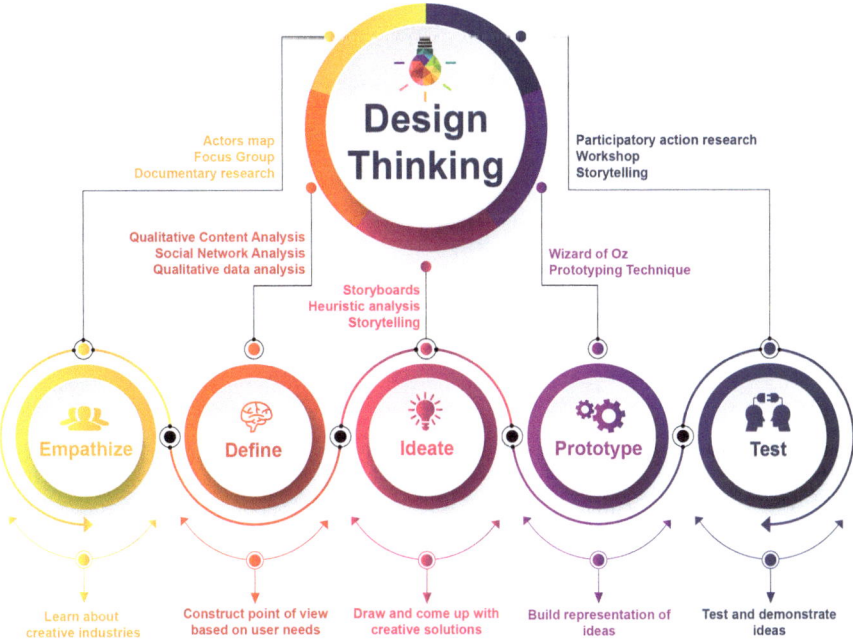

Fig. 3 Detail of the design thinking method used in the AiRT project. Source: Own elaboration, adapted from Both (2009)

Fig. 4 The parallel requirements to be satisfied in AiRT project methodology. Source: own elaboration

Fig. 5 Participatory process of the project. Source: Own elaboration

the software in accordance with this information (Step 4), and finally we tested the product with the same participants of the focus groups in a demonstration session, through the use of Participatory Action Research (McIntyre 2007) (Step 5).

The project began with the first step, in which a diagnosis of the needs of the creative industries was made and, after the analysis, a proposal of solutions was developed. To do so, different Service Design Tools were used. Firstly, the actors' map tool (Morelli 2007) was applied to represent who should be considered and their relations, in order to obtain a holistic and clear vision of the participants. At the same time, an analysis of ethical issues and risk analysis was carried out in order to obtain a global vision of the regulatory, safety and ethical constraints, and their causes and effects in order to overcome them with the main objective proposal, and fulfil customers' needs by accomplishing liability, technical feasibility and reliability.

The first phase of the project was called "Empathise". Three focus groups were carried out in Spain, the United Kingdom and Belgium, from which we were able to obtain information from 13 different sectors of the creative industries (40% were drone pilots). Previously, we completed a study of the academic literature in order to identify already known barriers and facilitators; previous interesting issues to consider were deployed.

In the second phase "Define" we analysed the data using the Qualitative Content Analysis method (Mayring 2014), the Social Network Analysis, SNA method (Borgatti et al. 2009), in addition to manually coding and categorising the qualitative data. As a result, we obtained the needs analysis, the ethical problems and the risk analysis. The results obtained formed the basis for the design of the drone, the development of the European policy book (De-Miguel-Molina and Santamarina-Campos 2018) and the redefinition of the Exploitation plan (AiRT 2017).

Subsequently, from the synthesis of the obtained information in the focus group, together with the specifications included in the grand agreement, in the third phase "Ideate", written scripts were elaborated that were later transferred to graphic scripts (storyboards) (see Fig. 4). Those represented the use of the AiRT system in different creative scenarios, allowing communication of the main ideas in a clearer way. From the storyboards, the requirements were derived to obtain specific functionality to be implemented in the GCS software. Furthermore, from the storyboards and the heuristic evaluation (Ball and Bothma 2017) of 30 flight plan software programs, graphic design and usability aspects were defined in the GCS software. To conclude this stage, a storytelling of the history of the project was developed, with the aim of informing the creative industries and other sectors of the potential of the tool. The "Ideas" phase is the step in which the final design of the system was carried out under the objectives and conditions established in the previous phase. In this stage, both the adaptation of the indoor positioning system and the optimisation of the RPAS were developed in parallel, performing a continuous evaluation of each task.

Once the adaptation and optimisation of the parts was completed, the "Proto-type" phase began, in which the integration of all the components was carried out: the positioning system, the RPAS and the graphic user interface, allowing a first implementation of the solution, through prototype testing by all project partners. Finally, the "Test" phase was carried out, in which the final evaluation of the prototype was carried out in different relevant environments in Spain, the United Kingdom and Belgium, involving the creative industries in the demonstration process by using the Participatory Action Research tool, PAR (McIntyre 2007). As a conclusion to this stage, a storytelling of the process of developing the tool was carried out for its dissemination in social networks and product presentation. And with the aim of publicising and showing the capabilities of the AiRT system to new industrial sectors, the project was completed with a demonstrative work-shop designed to show the potential of the product. This phase allowed us to determine the efficiency with which the phases of the plan had been executed, and the effectiveness and efficiency of the methodologies used to cover the detected needs.

Thus, the work plan developed is characterised by the constant evaluation of the tasks and the active participation of the creative industries throughout the entire project process. This circumstance determined the need to address to the identification, the development and testing of the prototypes, which, in turn conditioned the qualitative methodology to be used. Therefore, for the identification, the focus group was proposed (Sanders and William 2002) and for the demonstrator the PAR approach was applied. The purpose of the use of this type of technique was to obtain relevant data from the key informants (creative experts) that allowed for the subsequent interpretation and analysis of the facts based on the experiences.

The use of these qualitative techniques as methodological tools is justified by the importance conferred in this project to the creative SMEs, as main protagonists and beneficiaries. Engaging these professionals from the pre-existing idea through to the generation of the final product makes sense because both the needs analysis and the demonstration were proposed from a different focus; from the classic study of the object we now turn to the subject as the object of study. This has allowed us to provide fundamental information for the improvement cycle, and, once the project is completed, helps to tackle the commercialisation phase in which the definitive equipment will be built in a successful manner.

Both techniques were based on a script that allowed guiding of the conversations and experiences towards the topics of greatest interest that are linked to photography and filming in confined spaces. The different informants were selected based on a series of criteria and based on a geographical selection. This selection responds to the need to delimit the study area in order to carry out a comparative analysis of the creative industries in different spatial and sectoral contexts. This procedure made it possible to relate the similarities and differences obtained in each particular case, and then to establish general conclusions that allowed, in an objective way, not only the identification of the needs of each activity, but also the ability to approach a critical analysis and demonstration of the prototype with the active participation of the groups involved in the creative industries.

At the same time, the dissemination and exploitation of all the activities described above was focused on ensuring the dissemination and exploitation of the project results in an appropriate manner, facilitating a future commercialisation phase as well as the management of rights and patents among members of the consortium.

5   The Workplan and the Main Achievements

The workplan is composed of seven workpackages (see Fig. 6). The first one, called Management, is led by the UPV. The second one, Analysis of needs, security and risk, is Clearhead's responsibility. The third one, Adaptation and Integration of indoor positioning system, is Pozyx's task. The fourth one, RPAS design (which is already completed), and the fifth one, Integration, Validation and Demonstration, have been developed by AeroTools. The sixth one, Dissemination, is led by the UPV. And the last one, Exploitation Strategy, is led by AeroTools.

Regarding workpackage one, the biggest achievement we had is accomplishing lean management by establishing open communication, even if the consortium is interdisciplinary (see Fig. 7).

The objective has been achieved due to two fundamental aspects: on one hand, the use of the platform Basecamp[6] and, on the other one, the implementation of weekly technical meetings and collaborative and interdisciplinary work sessions, with partial and weekly deadlines for the final product.

Workpackage 2 began with the three focus group activities mentioned above and included the development of the needs analysis, the ethical issues and the risk analysis. Furthermore, the information retrieved was useful for the design of the drone, the GUI (graphical user interface) and the development of advanced functionalities.

In addition, this information was key for the development of the European policy book and to enrich the Exploitation plan. The main results of this work package are included in the third chapter "Creative industries' needs: a latent demand" and in the

---

[6]https://basecamp.com/

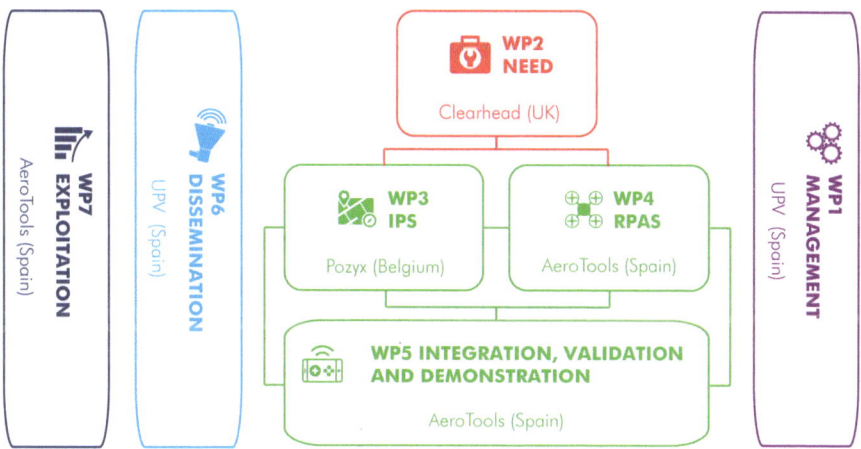

Fig. 6 Pert diagram of AiRT project. Source: Own elaboration

Fig. 7 Lean management. Source: Own elaboration

eighth chapter "The relationship of the industry with the public administration: co-regulation for training", where the main results obtained from the analysis of the needs, the ethical problems and the risk analysis of the creative industries are exposed.

In workpackage 3 we identified and implemented safety improvements for the indoor positioning system. We also developed methods to set up the system easily, including anchor discovery and auto-calibration. We have even implemented a four-antenna tag, which can be mounted on a drone. Furthermore, we developed customised algorithms for positioning, making optimal use of the four-antenna configuration. And finally, we developed a virtual environment map. The analysis and potential of this innovative indoor positioning tool will be tackled in the chapter titled "How cutting-edge technology can benefit the creative industries. The positioning system at work", where new functionalities applicable to the creative industries sector will be analysed.

In workpackage 4 the drone was redesigned in accordance to all the needs of the creative industries and the specifications in Annex 1 (part A) of the DoA,[7] in the Grant Agreement no 732,433. As a result, it is safe and secure, allowing high-quality filming and optimal compromise between drone size and payload. The main challenge reached in this work package will be illustrated in the ninth chapter, titled "Indoor drones for the creative industries: distinctive features/opportunities", through an analysis of the market opportunities of indoor drones in the sector of the creative industries.

In workpackage 5, in order to integrate all system components, the communication channels between different components have been established. An Intelligent Flight Control System was developed to control the RPAS remotely. This system is composed of the ground control system and the on-board control system and it is based on a mapping building system. All these systems can be accessed by means of the GUI.

The main methodological tool used for the development of these tasks was based on the use of storyboards, with the aim of facilitating the definition and summary of the functionalities of the AiRT system for its implementation (see example in Fig. 8). The results obtained from the use of this graphical tool will be collected in chapter seven, "Storyboarding as a means of requirements elicitation and user interface design. An application to the drones' industry".

Workpackage 6 is dedicated to the dissemination of the project results. More than 1100 people follow the progress of the project on Facebook. This was achieved through our website update activities, our work on social media, including both the professional sites (LinkedIn, Vimeo and Researchgate) and the transversal sites (known networks Facebook, Twitter, Instagram and YouTube). Moreover, project visibility is obtained through our participation in conferences, as well as the publication of several academic chapter books.

In workpackage 7, the Exploitation plan was revised by our focus groups and was complemented by an analysis of our competitors in the market. With this information and the value proposition of the prototype, AeroTools will be able to define more precisely the Exploitation plan. In chapter seven, "Indoor drones for the creative industries: distinctive features/opportunities", not only will the main innovations derived from the work package four be tackled, but also the market opportunities of the indoor drones in the sector of creative industries.

---

[7]Description of the action.

Fig. 8 Detail of the graphic script (storyboard UC00). Source: Own elaboration

6 Conclusion

The creative industry sector is composed mainly of SMEs. These companies find two main barriers when competing with bigger organisations: (a) obtaining economic resources for investing and, (b) being aware of the newest technological developments and having access to them.

The main objective of the European Commission Innovation Actions is to support the technological transference to the creative industries in order to increase their competitiveness by opening new markets and their set of final products and services (European Commission 2010).

In this chapter, we have summarised the main objectives, steps and achievements that the interdisciplinary European team has accomplished, fulfilling the European Commission requests. The European Commission innovative actions are short projects devoted to transfer knowledge and technology from university and specialised companies to society by means of team work and the search for a highly applied final product or service that benefits a maximum number of European SMEs.

From the Universitat Politècnica de València, the partner that manages the project, and considering society as our main stakeholder and know-how recipient, we are taking the lead on disseminating results by explaining the different aspects of the project at conferences, workshops and in an academic book with the aim of allowing companies and all interested members of the public to interact and discuss not only technical, but any other concern related to the final product and the whole project.

The dissemination actions have been complemented with a demonstration session seeking to go one step further, from the theoretical side to the hands-on activity. In this activity, only a reduced number of experts will participate. The benefits and high standard features of the indoor drone were tested in an incomparable site, the Principe Felipe Museum, located at the Valencia Arts and Science City, verifying the high level of acceptance that the product has among SMEs and interested stakeholders.

Annex: Focus Group Informants

Informant	Profile	Drone operator	Sectors	Date
1	TV/music/movies/ performing arts/publishing	No	Publicity, TV, cinema and videogames sectors	February 2017
6	Design/architecture	No	Architecture, museums and heritage sectors	February 2017
8	Antiques and museums/arts &crafts	No	Architecture, museums and heritage sectors	February 2017
13	Advertising/architecture/ photo	Yes	Architecture, museums and heritage sectors	February 2017
14	Museums/architecture	No	Architecture, museums and heritage sectors	February 2017
19	Movies/advertising	No	Publicity, TV, cinema and videogames sectors	February 2017
20	Photo/fashion/advertising/ music	No	Publicity, TV, cinema and videogames sectors	February 2017

References

AiRT (2017) Deliverable D7.1 AiRT exploitation and business plan, IPR strategy and agreement. Non-public deliverable

Ball L, Bothma T (2017) The importance of usability evaluation when developing digital tools for a library—a case study. In: Arvola P, Hintsanen T, Kari S, Kolehma S, Luolin S, Sillanpää J (eds) Improving quality of life through information. Proceedings of the XXV Bobcatsss symposium. University of Tampere, Tampere, pp 137–142

Borgatti SP, Mehra A, Brass DJ, Labianca G (2009) Network analysis in the social sciences. Science 323(5916):892–895

Both T (2009) Bootcamp Bootleg. https://dschool.stanford.edu/resources/the-bootcamp-bootleg

Brown T (2009) Change by design. How design thinking transforms organizations and inspires innovation. Harper Collins, New York

De-Miguel-Molina M, Santamarina-Campos V (2018) Ethics and civil drones. European policies and proposals for the industry. Springer, Germany

De-Miguel-Molina B, Segarra-Oña M (2017) The drone sector in Europe. In: De Miguel Molina M, Santamarina-Campos V (eds) Ethics and civil drones. European policies and proposals for the industry. Springer, Germany, pp 7–34

Droneii (2018) The drone market environment map 2018. Available via DRONEII. https://www.droneii.com/drone-market-environment-map-2018. Accessed 25 Jan 2018

European Commission (2010) Green paper: unlocking the potential of cultural and creative industries, COM/2010/0183 final, Brussels. http://eur-lex.europa.eu/legal-content/EN/TXT/PDF/?uri=CELEX:52010DC0183&from=EN. Accessed 15 Mar 2018

Johansson-Sköldberg U, Woodilla J, Çetinkaya M (2013) Design thinking: past, present and possible futures. Creat Innov Manag 22(2):121–146

Martin RL (2009) The design of business: why design thinking is the next competitive advantage. Harvard Business Press, Boston

Mayring P (2014) Qualitative content analysis: theoretical foundation, basic procedures and software solution. Available via SSOAR. http://nbn-resolving.de/urn:nbn:de:0168-ssoar-395173. Accessed 15 Mar 2018

McIntyre A (2007) Participatory action research, qualitative research methods series, vol 52. Sage, Thousand Oaks

Morelli N (2007) New representation techniques for designing in a systemic perspective. In: Paper presented at design inquires. University of Arts, Crafts and Design, Stockholm, 27–30 May 2007

Osterwalder and Pigneur (2010) Business model generation. Wiley, Canada

Sanders E, William C (2002) Harnessing people's creativity: ideation and expression through visual communication. In: Langford J, McDonagh-Philp D (eds) Focus groups: supporting effective product development. Taylor and Francis, London, p 137

Santamarina-Campos V (2017) EU policies about drones. In: De Miguel Molina M, Santamarina-Campos V (eds) Ethics and civil drones. European policies and proposals for the industry. Springer, Germany, pp 35–42

Stickdorn M, Schneider J, Andrews K, Lawrence A (2011) This is service design thinking: basics, tools, cases, vol 1. Wiley, Hoboken

The Economic Impact of the Creative Industry in the European Union

Rafael Boix-Domènech and Pau Rausell-Köster

Abstract This work analyses the economic impact of the creative industry in the European Union. The paper quantifies the direct and indirect impacts of the creative industry, concluding that they not only have a direct impact on the employment and the production, but also contributes to the technological progress and long-term development of the European Union. Most of this contribution is due to the creative service industries, whereas the direct contribution of the creative manufacturing industries is smaller.

1 Introduction

Now is the time to admit that the relationships between the creative ecosystem and the economic model are much more sophisticated than we previously thought and that they are connected in ways that go way beyond market exchanges.

In this paper, we ask what the impact is of the creative industries on the European Union (EU) economy. The creative industries contribute more than many traditional industries to the macroeconomic indicators of the EU, such as the production or the employment. However, the hypothesis we explore is that impact of the creative industries in the UE economy goes further than being a mere contribution to employment and production, and that they contribute to the technological progress and long-term development of the EU.

The role of the creative industries as a driver of innovation and a catalyst for economic transformation has become increasingly important in European regional policy. The new European regional policy (Foray et al. 2012) notes that the creative industries and, in particular, the creative services are in a strategic position to promote smart, sustainable and inclusive growth in all EU regions and cities, and

R. Boix-Domènech (✉)
Department of Economic Structure, Universitat de València, València, Spain
e-mail: Rafael.boix@uv.es

P. Rausell-Köster
Department of Applied Economics, Universitat de València, València, Spain

© The Author(s) 2018   19
V. Santamarina-Campos, M. Segarra-Oña (eds.), *Drones and the Creative Industry*,
https://doi.org/10.1007/978-3-319-95261-1_2

thus contribute fully to the Europe 2020 Strategy. The Research and Innovation Strategies (RIS3) guide argues that creative industries have multiple roles to play in unlocking the creative and innovative potential of a region (Foray et al. 2012). Also, the academic literature advocates the promotion of a sustainable and endogenous way of *resetting* an economy through a growth agenda that includes a role for creative industries in regional and local contexts (Cooke and De Propris 2011).

But the relations between cultural and creative sectors and the economic model go much further. The cultural field produces values, and values are one of the elements that determine our behaviour and govern the way we perceive the world. In fact, it is our set of values that sets the objectives of the institutions we create in order to articulate our life in society. The field of culture externalises values that permeate into the socio-economic space and seem to be much more in line with the concept of sustainable development. In terms of Ingleharts's *Cultural Evolution*, increasing emphasis in Posmaterialism values and Self-expression values (Inglehart 2018). Values emerging from the field of culture reflect a new hierarchy that includes aspects like the explicit wish to innovate, relational consumerism (as opposed to transactional consumerism), free exchange, critical thinking, personal development, solidarity, cooperation, networking, the value of diversity and beauty, participation and the importance of the recreational and vital dimension, as opposed to the purely economic gain. In other words, the actions of creativity are not exclusively guided by instrumental rationality. Expressive values and values of exchange and mutual benefit are also at work.

The goal of this paper is to provide a critical review of the literature about the direct and total impacts of the creative industries in the economy of the EU. The paper contributes to the literature by providing the most complete review of these effects elaborated to date, and by comparing the size of the direct effects with the total effects (direct, indirect and induced).

The paper is divided into five parts. Section 2 reviews the definition and measurement of the creative industries. Section 3 focuses on the scenarios and theoretical models to explore the economic impact of the creative industries. Section 4 reviews the direct and total effects of the creative industries on the EU economy in the literature, and how they change regarding the definition of creative industries used and the statistical models used for their measurement. Finally, Sect. 5 summarises and discusses the results.

2   The Creative Industries: Definition and Measurement

2.1   Concepts

The term *creative industries* originates in Australia (DCA 1994), and then its use expanded to the United Kingdom, which needed to find new bases for growth for its post-industrial economy (DCMS 1998; O'Connor 2007). They are also frequently referred to as *cultural and creative industries*, the *creative industry*, or the *creative sector*, even if some authors introduce nuances in the use of one or the other term.

There are many definitions of creative industries. The most usually cited are those by the British Department of Culture Media and Sports (DCMS), the United Nations Conference for Trade and Development (UNCTAD), and the European Union authorities (European Commission, European Parliament):

(a) The DCMS (2001: 5) refers to "industries which have their origin in individual creativity, skill and talent and which have a potential for wealth and job creation through the generation and exploitation of intellectual property". Creative industries are signs of the natural evolution of the cultural industry that follow the structural changes caused by the affirmation of new technologies and new products in the sphere of the entertainment industry.

(b) UNCTAD (2008: 4) defines creative industries as "cycles of creation, production and distribution of goods and services that use creativity and intellectual capital as primary inputs; constitute a set of knowledge-based activities, focused on but not limited to arts, potentially generating revenues from trade and intellectual property rights; comprise tangible products and intangible intellectual or artistic services with creative content, economic value and market objectives; are at the cross-road among the artisan, services and industrial sectors; and constitute a new dynamic sector in the world trade". The term "creative industries" exceeds the limits of the cultural sector to include media and ICTs following the structural changes due to the growth and development of the new technologies.

(c) For the European Parliament (2016: 10) they are defined as "those industries that are based on cultural values, cultural diversity, individual and/or collective creativity, skills and talent with the potential to generate innovation, wealth and jobs through the creation of social and economic value, in particular from intellectual property".

2.2 Measurement

The concept of creative industries is made operational by drawing up a list of activities in which creativity is particularly important. So far there is no agreement on which of the approaches to reach this list is the most appropriate or what activities should be classified as creative (UNCTAD 2008; Throsby 2008).

UNCTAD (2008, 2010) distinguishes five conceptual models for its definition:

(a) the *DCMS model*, based on individual creativity and that marks the relevance of technological creative industries in contrast to the traditional cultural industries;

(b) the *symbolic texts model* differentiates between central, peripheral and border cultural industries, and focuses on the processes through which thought culture and creativity are produced, disseminated and transmitted;

(c) the *concentric circle model* assumes that the creative capacity and content is different in each industry. Creative ideas originate in a core creative circle formed by the creative arts, which then diffuse outwards through a series of concentric circles in which the amount of creative content decreases successively: other core industries, less central industries, and related industries;

(d) the *copyright model* of World Intellectual Property Organization (WIPO) and International Intellectual Property Alliance (IIPA) distinguishes between industries that produce most of the intellectual property (core copyright industries), those necessary to convey the goods and services to the consumer (interdependent copyright industries), and those in which intellectual property is only a minor part (partial copyright industries);

(e) and the UNCTAD model, which attempts to classify creative industries into domains, groups and subsectors, distinguishing between heritage, arts, media and functional creations. This model differentiates between upstream activities (traditional cultural activities), and downstream activities (much closer to the market).

Other research has proposed to classify an industry as creative based on its share of creative occupations or/and the qualification in terms of educational levels of the employees (Higgs and Cunningham 2008; Nathan et al. 2015 for NESTA (2015)).

The choice of one or the other approach and the delimitation of the following list of sectors to measure the creative industries usually provokes controversy and not only a few arguments. The most comprehensive list seems to be the provided by UNCTAD (2010), including: printing (only for statistical comparisons where data are not enough disaggregated), publishing, advertising and related services, architecture and engineering, specialised trade of art in antique markets, crafts, fashion and high-end industries, specialised design services, film, motion and video, musing and sound recording, performing arts, other visual arts, photography, radio and television broadcasting, software and videogames, heritage and cultural occupations, copyright collecting societies, cultural tourism and creative research and development (R&D).

However, as can be seen in Table 1, other classifications do not consider industries such as printing, creative retail trade, crafts, designer fashion, heritage, interactive media, some visual arts, copyright societies, cultural tourism, recreational services, engineering and R&D services.

Since most of the industries are shared between the most usual classifications (see Throsby 2008), differences resulting from the application of one classification or another can be moderate, and their importance depends on the objective of the research and the places to which they are applied (see Lazzeretti et al. 2018). For example, Boix et al. (2013) and Boix and Soler (2017) argue that manufacturing sectors usually classified as creative are basically engaged in making more than in creating, and their inclusion biases the results of the estimation of the impacts of the creative industries, so that they should be better focused as *semi-creative industries* (Boix 2012).

Although the use of standard lists is useful in comparisons (in particular between cities, regions and countries), it is still debatable whether the same economic sector can be creative in one place and not in another. In addition, the classifications are

Table 1 Classifications of creative industries

Creative industries	DCMS (2009) (UK)	WIPO copyright industries (2003)	Eurostat LEG (2000)	KEA European Affairs (2006)	UNCTAD (2010)	European Parliament (2016)	NESTA (2015) (Nathan et al.)
Printing		X			X[a]		
Publishing	X	X	X	X	X	X	X
Advertising and related services	X	X	X	X	X	X	X
Architecture and engineering	X	X	X	X	X	X	X (only architecture)
Arts and antique markets/trade	X	X		X	X	X	
Crafts (artistic)	X	X	X	X	X	X	
Design/specialised design services	X	X	X	X	X	X	X
Designer fashion and high-end industries	X	X			X	X	
Film/motion picture and video industries	X	X	X	X	X	X	X
Music/sound recording industries	X	X	X	X	X	X	X
Performing arts (theatre, dance, opera, circus, festivals, live entertainment)/independent artists, writers, and performers	X	X	X	X	X	X	X
Photography	X	X	X	X	X	X	X
Radio and television (broadcasting)	X	X	X	X	X	X	X
Software, computer games and electronic publishing	X	X	X	X	X	X	X
Heritage/cultural sites (libraries and archives, museums, historic and heritage sites, other heritage institutions)			X	X	X	X	X
Interactive media			X	X		X	X
Other visual arts (painting, sculpture)			X		X	X	X
Copyright collecting societies				X			

(continued)

Table 1 (continued)

Creative industries	DCMS (2009) (UK)	WIPO copyright industries (2003)	Eurostat LEG (2000)	KEA European Affairs (2006)	UNCTAD (2010)	European Parliament (2016)	NESTA (2015) (Nathan et al.)
Cultural tourism/recreational services				×	×		
Creative R&D					×		
Public relations and communication activities/ translation and interpretation							×

Source: Own elaboration
[a]Only used for statistical reasons in comparisons

updated over time, trying to introduce more objective criteria for the delimitation of sectors, as discussed by Nathan et al. (2015).

2.3 European Statistics for the Creative Sectors

Due to the dual nature of the creative sector, the statistics for its measurement can focus on the purely cultural or purely technological part when a restricted version of creativity is adopted.

Focusing on a more *cultural* approach, a precursor is the "Framework for Culture Statistics" (FCS), adopted at the Conference of European Statisticians of 1986 (see UNESCO 2009). The guiding principles included in the 2009 FCS are aimed at establishing a conceptual foundation that encompasses all cultural expressions and uses internationally comparable categories for the classification of products, economic activities and occupations, facilitating the identification of variables and the elaboration of cultural indicators to capture the cultural reality in a way that is both descriptive and integrating (Coll et al. 2018). The guidelines were applied to the elaboration of world cultural statistics in UNESCO (2012).

An increase in the level of awareness about the lack of cultural statistics in the European Union (EU) in the context of several international forums led to the establishment of the European Leadership Group on Cultural Statistics (LEG-Cultura) in 1997. The main goal of the LEG was to define the cultural *domains*, suggest changes to national and international classifications to reflect the distinctive features of the cultural and creative sector, revise and carry out an inventory of all the available data and create a series of indicators that allowed comparison between nations.

In 2009, Eurostat proposed the creation of the European Statistical Network (ESSnet-Culture) to improve the methodology and production of data about cultural sectors, and at the same time facilitate their comparison at the European level. The results were published in the form of a *Cultural Statistics* pocketbook (2007, 2011 and 2016) as part of the *Statistical Books* series. This was the first publication by Eurostat related to the field of culture that includes comparable data available in the European Statistical System and other sources of information such as UNESCO or the Eurobarometer. The data included in the Eurostat's Culture Statistics report and their original sources are: cultural employment (EU Labour Force Survey), cultural enterprises (Business Statistics and Business Demography), international trade in cultural goods (Eurostat), cultural participation (Adult Education Survey by Eurostat), use of ICT for cultural purpose (Community survey on ICT usage in households), and private spending in culture (Household Budget Survey).

At the opposite extreme, we find the reports and statistics elaborated by WIPO (2014) and the European Union Intellectual Property Office (EUIPO 2016), which focus on the *intellectual property rights* (IPR) part of the creative industries. This includes industries that register trademarks, patents, designs, copyright, geographical indications and plant variety rights, which means that some cultural industries (e.g. those related to heritage) are not considered. Both extremes—culture-based and IPR-based industries—have a part in common, and consider core definitions and enhanced perimeters.

Between these two extremes, other databases containing information on creative industries in the EU have been drawn up, using Eurostat or other public institutions as the original source of data. These include the European Cluster Observatory (ECO) (http://www.clusterobservatory.eu/) and the Toolkit of Creative Med (ToCM) (http://toolkit.creativemed.eu/). The ECO database has information for the period 1991–2011 (the quality and coverage of the information is uneven between years), at different territorial levels (NUTS 0 to NUTS 2), and considers as creative and cultural industries: advertising, artistic and literary creation, preservation of historical sites and buildings, other printing and publishing, radio and television, creative retail and distribution and software.

The ToCM is oriented towards the elaboration of policies based on creativity in the Mediterranean area of the EU, although the database contains information for all of the EU. The database is at NUTS 2 level and mainly refers to the year 2011. Data on workers in cultural and creative industries are taken from the ECO, although ToCM also includes indicators of semi-creative industries, creative class and cultural resources. It is already capable of automatically generating diagnoses and suggesting policies for each region, within the framework of the strategy RIS3.

Despite the effort made in recent times to improve the quantity and quality of creative industries' statistics, there is no doubt that more sophisticated measurement systems are needed if we are to gauge and analyse all the sophisticated relationships between the creative industries and their economic and social impacts.

3   The Economic Impact of the Creative Industries: Scenarios and Theoretical Models

3.1   Scenarios

The economic impact of the creative industries on the European Union can be analysed in terms of *direct* contribution to the production, employment, productivity, innovation, entrepreneurship, etc. In addition to this contribution, the most important effect of the creative industries can be found in its *indirect* effects, that is, its multiplying effects on the economy. In addition, other two issues should be taken into account: first, that these direct and indirect effects could be positive or negative; and second, that the intensity and direction of the effect could be different between locations.

General scenarios relating the creative industries to economic performance have been proposed by Potts and Cunningham (2008). They propose four scenarios: welfare, competitive, growth and innovation:

(a) In the *welfare model*, the creative industries are deemed to be affected by Baumol's disease (Baumol and Bowen 1965) and their productivity grows at a slower rate than does the rest of the economy, even if they are subsidised. This is because in this model the creative industries are oriented to the enhancement of welfare.

(b) In the *competitive model*, the creative industries are seen as just another industry and have no more effect on technological change, innovation or productivity than do any other activities.

(c) In the *growth model*, the creative industries are said to be a growth driver and their impact on the economy is more than proportional and caused by both demand and supply influences. Demand influences occur when production or income growth results in increased demand for creative services, which, in turn, feed into and change production, or incomes, multiplying final effects (Potts and Cunningham 2008; Rausell et al., 2011). Supply-side influences can be conceptualised as occurring through five basic mechanisms through which creative industries act as a growth driver:

(c.1) First, there is a sectoral effect (Potts 2009; Lee 2014): creative industries have been growing more than other industries simply because they have been in an expansion phase, caused by increased investment and qualitative improvement in supply and input factors, institutional change and intra-sector externalities.

(c.2) Second, there is an industry spillover through supply-chain linkages into other sectors (Bakshi and McVittie 2009; UNCTAD 2010; Lee 2014).

(c.3) Third, there is the effect of improving the technical efficiency of the aggregate economy (Yu et al. 2014).

(c.4) Fourth, a knowledge spillover effect (Potts and Cunningham 2008; Bakshi and McVittie 2009; Chapain et al. 2010) occurs thanks to the introduction of new ideas that spread to other sectors, acting as a catalyst for the generation of innovations, and improving the absorptive capability of innovations in other sectors.

(c.5) And, finally, there is the supply-side effect of creating an amenity value that attracts skilled workers and tourists (Lee 2014).

(d) Finally, in the *innovation model* the creative industries are seen as being a part of a process of economic evolution and their role is to provide *evolutionary services* to the innovation system, facilitating change of the entire economic system (see Potts 2009).

The innovation model is supported by the theory of differentiated knowledge bases (Asheim et al. 2011; Asheim and Parrilli 2012). Analytical and synthetic knowledge bases are well-known in economics: the term analytical knowledge base refers to the development of new knowledge through the use of the deductive scientific method and scientific laws (e.g. the pharmaceutical industry has an analytical knowledge base), and the term synthetic knowledge base refers to the generation of knowledge by an inductive process of testing, experimentation, and practical work (e.g. the mechanical engineering industry develops on a synthetic knowledge base).

Of particular interest has been the introduction of a third category, the symbolic knowledge base. This type is normal for the cultural and creative industries in that it refers to the "creation of meaning and desire as well as aesthetic attributes of products, producing designs, images and symbols, and to the economic use of such forms of cultural artefacts" (Asheim et al. 2011). In the case of the symbolic base, knowledge inputs and outputs are aesthetic more than cognitive, and new knowledge is usually developed through a creative process rather than through analytical or problem-solving processes. Creative industries provide services to the rest of the productive system in two ways: as inputs to other industries, and also via a horizontal spillover effect on the perceptions of people, businesses and institutions.

Note that in these four scenarios the effects of the creative industries are basically neutral or have some positive aspect. A fifth scenario can be added, in which the effects of the creative industries on wealth and welfare are negative. This effect could be due to several causes, among them:

(a) a crowding out effect of the creative industries on other economic activities, in the case that the creative industries do not have significant impacts on welfare, or this effect exceed the initial increases in welfare considered by Baumol and Bowen (1965);
(b) the precariousness of the labour relations model that seems to accompany this type of industry (Hesmondhalgh, 2010);
(c) or effects of trivialisation, alienation or propaganda as initially considered by Horkheimer and Adorno in the *Dialectic of Enlightenment* (2002), and that produced a long-term joint crowding out effect on wealth and a reduction of freedoms.

3.2   Theoretical Models

Sacco and Segre (2006), Rausell et al. (2011), Marco et al. (2014) and Boix and Soler (2017) have developed analytical models that relate the creative industries with wealth, productivity and well-being.

Sacco and Segre (2006) propose a virtuous circle based on the acquisition of competences, where the notion of competence refers to the effect of the stimulus of cultural, symbolic and identitarian capital. The basic assumption is that the level of competence and capability of consumers (some of which are creative workers) is sufficient enough to guarantee that they would be willing to pay extra for the creative component of a product. In order to meet this demand, firms invest in the skills of creative workers in order to increase the creative component in the enterprise's goods and services. The result is an increase in a locality's stock of creative capital. Such changes in local cultural supply and rising social awareness lead to improvements in the competences of non-core creative workers while fostering demand for creative commodities. At this point, a part of the value added generated by the process is then devoted by firms to financing creative activities and by the public sector in investing in creative industries, creating a virtuous cycle.

Rausell et al. (2011) and Marco et al. (2014) propose a theoretical framework with circular causal effects: an increase in GDP per capita increases the share of people with high levels of education and income, the percentage of public and private expenditure oriented to creative goods and services and the stock of cultural capital. The result is an increase in the demand for creative goods and services that then engenders a growth in the share of workers in the creative industries. This has two effects. First, there is an increase in an economy's overall number of innovations due to, on the one hand, the addition of those innovations produced by the creative industries (supply side), and, on the other hand, a higher propensity to consume innovations by workers employed in the creative industries (demand side). Second, there is an increase in the level of productivity of the economy as a whole, under the assumption that productivity in the creative industries is higher than the average for the economy. Increases in innovation and productivity result in an increase in GDP per capita, and so the process starts again.

Boix and Soler (2017) address the lack of robust analytical modelling of previous research on the effects of creative industries and use a semi-endogenous technological change model to analyse the relationship between creative industries and regional productivity. The model assumes that the creative industries lead to an increase in product variety that raises productivity by allowing the spread of intermediate production more thinly across a larger number of activities, each of which is subject to diminishing returns and hence exhibits a higher average output when operated at a lower intensity. The implication is that the way to increase productivity levels is by dedicating a large fraction of output to creative activities.

4 Measuring the Economic Impact of the Creative Industries in the European Union

As explained in Sect. 3, the empirical research that has measured the impacts of creative industries has focused on their direct and total effects on production, income, wages, employment, productivity or innovation. The size of this effect depends to a great extent on which sectors have been included in the empirical definition of the creative industries.

4.1 Direct Impact of the Creative Industries

The initial reports by the Department of Culture, Media and Sports stated that at the end of the twentieth century, the creative industries were bringing about 8% of national income, employing 5% of the workforce in the United Kingdom and growing at 8% per year.

Reports based on a culture-based approach (e.g. KEA 2006; Power and Nielsén 2010) and including both manufacturing and service creative industries, stated that the size of the creative sector is about 2.6% of the EU Gross Domestic Product (GDP) and employment (from 4.7 to 6.7 million employees, depending on the countries included in each report), with a turnover of 654,000 million € and growth rates of the GDP and employment of more than 4% annual average.

Later studies based on more comprehensive and services-oriented definitions of the creative sector suggest that their contribution to the economy is substantially greater. TERA (2010, 2012) stated that core creative industries accounted for 3.8% of the workforce and 4.5% of the value added in the EU, and core plus non-core creative industries were about 6.5% of the employment (14.4 million employees in 2008 and 14 million in 2011) and 6.9% of the EU value added. EY (2014) stated a contribution of the creative industries to the EU economy of 535.9 billion €, 4.2% of the GDP and 3.3% of the EU's active population (7.1 million employees). These reports also stated that until 2007, the creative industries were a fast-growing sector, and that after 2008 they withstood the economic crisis.

A recent research by Boix and Soler (2017) measured the size of the creative industries for 250 regions in 24 countries of the EU, finding that creative industries generated 7.8% of the total production (GDP), 7.9% of total employment, and that their labour productivity was 1.2% lower than the European average. The authors find important differences between manufacturing and service creative industries, basically a low contribution and impact of creative manufacturing and a high contribution and impact of creative services. Thus, creative manufacturing was 1% of the GDP and 1.6% of the employment, and their productivity was 41% below the European average. Creative services were 6.8% of the GDP and 6.2% of the employment, and their productivity was 9% higher than the European average.

All this evidence is consistent with research for outside Europe. For example, Dolfman et al. (2007) found that in the United States the average wage in the creative industries was 34.9% higher than the national average private sector wage. Potts and Cunningham (2008) provided evidence that in Australia the average incomes of workers in the creative industries were 31% higher than national average incomes, and that the aggregate growth rate of creative industry incomes was higher than incomes for the aggregate economy.

4.2 Spillovers and Total Effects of the Creative Industries

There are several ways to measure the total effect of the creative industries in the economies. A traditional way, widely used in the economy of culture, has been the use of input-output tables and multipliers (Throsby 2008), although this approach has not yet been applied to the EU as a whole.

Another way is the use of econometric models, which have been used to assess the impact of the creative industries in the economy of a single country (Rausell et al. 2011; Lee 2014) or in the regions of the EU (De-Miguel-Molina et al. 2012; Marco et al. 2014; Boix et al. 2013; Boix and Soler 2017; Boix and Peiró 2017) (see Table 2).

For a single country, Rausell et al. (2011) find that an increase of 100% in the share of creative industries is associated with an increase in GDP per capita of the Spanish regions of about 40%, although their estimates did not take account of some controls. Lee (2014) finds that in the UK travel-to-work areas, the creative industries are associated with positive differences in wages relative to other sectors of the economy, showing a relative effect of between 4.7% and 6.6% for each 100% increase of employment in the creative industries.

For a wider sample covering most of the European regions, De-Miguel-Molina et al. (2012), Marco et al. (2014) and Boix et al. (2013) found that the relative effect of the creative industries on the differences in GDP per capita was between 39% and 44%. However, Boix and Soler (2017) observed that these high elasticities were due to the misspecification of the empirical models, and departing from a theoretically robust semi-endogenous growth model amended the results of previous research on European regions. They found that the effect of creative industries on productivity was much lower than previously thought, with elasticities from 4% to 13%. These results were confirmed by Boix and Peiró (2017).

In a similar range are the results by Hong et al. (2014) for the provinces of China. The authors model the effect of agglomeration of creative industries (using a location quotient for creative industries) on total factor productivity (TFP) growth, finding that an increase of 100% in specialisation in creative industries increases TFP by about 4%. They also find evidence that the whole effect is due to the impact of creative industries on technological progress, whereas their effect on changes in technical efficiency was not significant. Boix and Soler (2017) also provided evidence that about 90% of the impact of the creative industries on the productivity of the EU regions was indirect, due to their spillover effects on the economy.

Some of these works discuss the implications or sensibility of the results to the empirical definition of creative industry and, in particular, the effect of the different creative industries on the productivity or the wealth. Boix et al. (2013) and Boix and Soler (2017) find robust evidence that only creative service industries have a positive and relevant effect on the differences of wealth between the regions of the European Union, whereas the aggregated effect of the so-called creative manufacturing industries tends to be null or negative.

Boix et al. (2013) disaggregated the correlations of individual creative sectors with the wealth of the European regions, finding that the correlation is positive for all the creative sectors, being particularly high for Publishing, Computer programming, Advertising, Architecture and engineering, creative Research and development and Creative retail. The correlation is lower for Broadcasting, Design and photography, and Arts, entertainment and recreation.

Table 2 Total effects of the creative industries in the GDP per capita, productivity and wages in econometric research

Article	Approach	Application	Target	Explanatory variable	Controls	Relative effects (elasticities) to a 100% change in creative industries
Rausell et al. (2011)	Endogenous virtuous circle / VAR levels/logs	Spanish regions 2000–2008	GDP per capita	Percentage of persons employed in cultural sectors	No	44%
De-Miguel-Molina et al. (2012)	Empirical modelling / Linear equation in levels	European regions 2008	GDP per capita	Percentage of persons employed in creative industries	Industrial structure, localisation economies	45%
Boix et al. (2013)	Empirical modelling / Linear equation in levels	European regions 2008	GDP per capita	Percentage of persons employed in creative industries	Industrial structure, localisation and urbanisation economies	39%
Marco et al. (2014)	Virtuous circle / Structural equations model in levels (dynamic)	European regions 1999–2008	GDP per capita; per capita disposable household income; labour productivity	Percentage of persons employed in creative industries	Urbanisation economies, higher education	41%
Lee (2014)	Labour market equation and fixed effects	UK travel-to-work areas 2003–2008	Hourly wage in non-creative industries (total and private)	Total creative industries employment	Industrial structure, population, skill, gender, age, migrants	4.7–6.6%
Hong et al. (2014)	Total factor productivity and fixed effects	Chinese provinces 2003–2010	Total factor productivity	Location quotient of employees in creative industries	Industrial structure, FDI, firm size	4%
Boix and Soler (2017) and Boix and Peiró (2017)	Semi-endogenous growth model log-linear	European regions 2008	Labour productivity	Percentage of persons employed in creative industries. Division between creative manufacturing and creative services	Industrial structure, MAR-Jacobs-porter economies, capital investment, network economies, innovation, human capital, creative class, spatial dependence	4–13%

Source: Own elaboration

5   Conclusions

This paper focuses on the economic impact of the creative industries in the economy of the European Union, differentiating between direct and total effects. The review of the literature measuring both types of impacts suggests the following conclusions:

(a) The creative industries have relevant direct and indirect effects in the economy of the EU (see KEA 2006; Power and Nielsén 2010; TERA 2010, 2012; EY 2014; Boix and Soler 2017). They have a direct contribution to employment, which ranges from 2.6% of the production and employment in the most conservative delimitations based on a culture-based approach, to 7.8% of the production and employment using comprehensive delimitations, such as those based on UNCTAD (2008, 2010).

(b) However, the creative industries not only have direct effects but also generate spillovers that affect the rest of the European economy through indirect, induced and evolutionary mechanisms (see Rausell et al. 2011; De-Miguel-Molina et al. 2012; Marco et al. 2014; Boix et al. 2013; Boix and Soler 2017). These mechanisms include supply and demand effects on production growth, and in particular, indirect effects related to the capacity of the creative industries to create technical progress and modify the evolutionary growth path of the economy. The initial measurement of these effects resulted in extremely high responses of the EU economy—elasticities of more than 40%—to increases in the contribution of creative industries to employment. However, these elasticities were hard to believe, and later research using more robust theoretical models and controlling the specification biases has proven that the response elasticities would be in a range of between 4 and 13%, which continues to be a high impact and deserves the attention of economic policies.

(c) Most of this contribution is due to the creative service industries, whereas the direct contribution of the creative manufacturing industries is small and the total contribution, including direct and indirect effects, could be negative. The direct contribution of the creative manufacturing industries does not exceed 1% of the GDP and 1.6% of the employment. Labour productivity of the creative manufacturing sector is about half of the EU average, whereas labour productivity of creative services is slightly higher than the EU average. The econometric estimates by Boix et al. (2013) and Boix and Soler (2017) show that higher levels of creative manufacturing reduce the labour productivity of the regions, whereas higher levels of creative service industries have a multiplying effect on the regional labour productivity in the EU.

Due to space limitations, this paper has not focused on the direct measurement of the relationship between creative industries and other variables such as innovation, foreign direct investment and exports of goods and services (especially tourism). In any case, the evidence at the European level about both is still small and is one of the fields in which more research is needed.

In fact, one final conclusion of this paper would be that the evidence on the effects of the creative industries on the EU economy is still very limited and needs much more research. We suggest some of the priority research lines. First, the measurement of the differences in economic impact of the creative industries for Mediterranean, Eastern, Central and Nordic regions needs study, as does the comparison with the effects for non-European countries. Second, research is needed on the measurement of the effects of the creative industries on the well-being of the EU economies, including their economic implications, to satisfy the cultural and creative rights of its citizens, which is the fastest way to enhance their utility through pleasure, engagement and meaning (see Rausell 2018). Third, we are lacking studies on the use of time-dynamic models, which is, however, limited by the current state of the statistics. Fourth, an approach that has been scarcely studied in the literature is the role of demand. The number of people employed in the cultural and creative sector determines the power of a solvent demand that is very prone to innovation and therefore becomes a promoter of social and political innovation. This is not only due to an income effect, as the *creative class* is also manifested through a particular lifestyle that involves consuming more innovative products and services and creative content. Fifth, in addition to the empirical measurements, we should move towards a complex and comprehensive conceptual model that considers all the relationships analysed.

References

Asheim BT, Parrilli MD (2012) Introduction: learning and interaction—drivers for innovation in current competitive markets. In: Asheim BT, Parrilli MD (eds) Interactive learning for innovation: a key driver within clusters and innovation systems. Palgrave Macmillan, Basingstoke, pp 1–32

Asheim B, Boschma R, Cooke P (2011) Constructing regional advantage: platform policies based on related variety and differentiated knowledge bases. Reg Stud 45:893–904

Bakshi H, McVittie E (2009) Creative supply-chain linkages and innovation: do the creative industries stimulate business innovation in the wider economy? Innov Manag Policy Pract 11:169–189

Baumol WJ, Bowen W (1965) On the performing arts: the anatomy of their economic problems. Am Econ Rev 55:495–502

Boix R (2012) Creative industries in Spain: the case of printing and publishing. In: Lazzeretti L (ed) Creative industries and innovation in Europe: concepts, measures and comparative case studies. Routledge, Abingdon, pp 65–85

Boix R, Peiró J (2017) Industrias de servicios creativos y productividad del trabajo en las regiones de la Unión Europea. In: Valdivia M, Cuadrado JR (eds) La economía de las actividades creativas: Una perspectiva desde España y Mexico, 1st edn. Universidad Nacional Autónoma de México and Universidad de Alcalá de Henares, Madrid, pp 357–380

Boix R, Soler V (2017) Creative service industries and regional productivity. Pap Reg Sci 96 (2):261–279

Boix R, De Miguel B, Hervás JL (2013) Creative service business and regional performance: evidence for the European regions. Serv Bus 7(3):381–398

Chapain C, Cooke P, Propris LD, MacNeill S, Mateos-García J (2010) Creative clusters and innovation. NESTA Research Report, London

Coll V, Pardo C, Caballer M (2018) Cultural statistics in Europe. In: Paper presented at the II forum in cultural economics. Two points in the silk road. Transmaking Spring Academy, University of Valencia, 5–8 Mar 2018

Cooke P, De Propris L (2011) A policy agenda for EU smart growth: the role of creative and cultural industries. Policy Stud 32(4):365–375

DCA (1994) Creative nation: Commonwealth cultural policy. DCA, Canberra

DCMS (1998) The creative industries mapping document. DCMS, London

DCMS (2001) Creative industries mapping document 2001. DCMS, London

DCMS (2009) Creative industries economic estimates statistical bulletin January 2009. DCMS, London

De-Miguel-Molina B, Hervás JL, Boix R, De-Miguel-Molina M (2012) The importance of creative industry agglomerations in explaining the wealth of European regions. Eur Plan Stud 20 (8):1263–1280

Dolfman ML, Holden RJ, Fortier Wasser S (2007) The economic impact of the creative arts industries: New York and Los Angeles. Month Lab Rev (Oct):21–34

EUIPO (2016) Intellectual property rights intensive industries and economic performance in the European Union. EUIPO, Munich

European Parliament (2016) Report on a coherent EU policy for cultural and creative industries (2016/2072(INI)). European Parliament, Brussels

EY (2014) Creating growth: measuring cultural and creative markets in the EU. EY, France

Foray D, Goddard J, Beldarrain XG, Landabaso M, McCann P, Morgan K and Ortega-Argilés R (2012) Guide to research and innovation strategies for smart specialisations (RIS 3). European Commission, DG Regional Policy

Hesmondhalgh D (2010) User-generated content, free labour and the cultural industries. Ephemera 10(3/4):267–284

Higgs P, Cunningham S (2008) Creative industries mapping: where have we come from and where are we going? Creat Industr J 1(1):7–30

Hong J, Yu W, Guo X, Zhao D (2014) Creative industries agglomeration, regional innovation and productivity growth in China. Chin Geogr Sci 24(2):258–268

Horkheimer M, Adorno TW (2002) Dialectic of enlightenment. Philosophical Stanford University Press, Stanford

Inglehart RF (2018) Cultural evolution: people's motivations are changing, and reshaping the world. Cambridge University Press, Cambridge

KEA (2006) The economy of culture in Europe. European Commission Directorate-General for Education and Culture, Brussels

Lazzeretti L, Boix R, Sánchez D (2018) Entrepreneurship and creative industries in developing and developed countries. In: Lazzeretti L, Vecco M (eds) Creative industries and entrepreneurship: paradigms in transition from a global perspective. Edward Elgar, Cheltenham, pp 35–57

Lee N (2014) The creative industries and urban economic growth in the UK. Environ Plan A 46:455–470

LEG Eurostat (2000) Cultural statistics in the EU, Eurostat working paper, population and social conditions series, 3/2000/E/No1, final report of the LEG. Eurostat, Luxembourg

Marco F, Rausell P, Abeledo R (2014) Economic development and the creative industries: a Mediterranean tale of causality. Creat Industr J 7(2):81–91

Nathan M, Pratt A, Rincón-Aznar A (2015) Creative economy employment in the EU and the UK: a comparative analysis. NESTA, London

O'Connor J (2007) The cultural and creative industries: a review of the literature. Arts Council England, London

Potts J (2009) Why creative industries matter to economic evolution. Econ Innov New Technol 18 (7–8):663–673

Potts J, Cunningham S (2008) Four models of the creative industries. Int J Cult Policy 14 (3):233–247

Power D, Nielsén T (2010) Priority sector report: creative and cultural industries. ECO, Stockholm

Rausell P (2018) Culture, creativity and economic progress. In: Mickov B, Doyle JE (eds) Culture, innovation and the economy, 1st edn. Routledge, Abingdon, pp 3–5

Rausell P, Marco F, Abeledo R (2011) Sector cultural y creativo y riqueza de las regiones: en busca de causalidades. Ekonomiaz 78:67–89

Sacco PL, Segre G (2006) Creativity, cultural investment and local development: a new theoretical framework for endogenous growth. In: Fratessi U, Senn L (eds) Growth and innovation in competitive regions: the role of internal and external connections. Springer, Berlin, pp 281–294

TERA Consultants (2010) Building a digital economy: the importance of saving jobs in the EU's creative industries. Tera, Paris

TERA Consultants (2012) The economic contribution of the creative industries to EU GDP and employment. Tera, Paris

Throsby D (2008) Modelling the cultural industries. Int J Cult Policy 14(3):217–232

UNCTAD (2008) Creative economy. Report 2008. UNCTAD, Geneva

UNCTAD (2010) Creative economy. Report 2010. UNCTAD, Geneva

UNESCO (2009) The 2009 UNESCO framework for cultural statistics (FCS). UNESCO, Geneva

UNESCO (2012) Measuring the economic contribution of cultural industries. UNESCO, Geneva

WIPO (2003) Guide on surveying the economic contribution of the copyright industries. WIPO, Geneva

WIPO (2014) WIPO studies on the economic contribution of the copyright industries. WIPO, Geneva

Yu W, Hong J, Zhu MD, Guo X (2014) Creative industry clusters, regional innovation and economic growth in China. Reg Sci Policy Pract 6(4):329–347

Creative Industries' Needs: A Latent Demand

Blanca de-Miguel-Molina and Marival Segarra-Oña

Abstract This chapter focuses on the needs of the creative industries and how the market for their services is increasing. For this purpose, the chapter is divided into two main analyses, the first covering the evolution of these sectors in the last years, while the second is focused on the advantages, advances and challenges that these industries are facing in relation to digitalisation and other emerging technologies. In the analysis of sectors, data about the number of enterprises, employees, turnover and value added were used. Concerning technologies influencing creative industries, a content analysis of 27 documents elaborated by leading consulting firms during the years 2016, 2017 and 2018 was elaborated. From this analysis, 160 codes were defined to express which advances, advantages and challenges these consulting firms have indicated for creative industries. These consulting firms are at the forefront, advising their clients on how to implement different emerging technologies, such as artificial intelligence, augmented and virtual reality and blockchain, among others.

1 Introduction

According to Culture Action Europe (CAE), the biggest European cultural network, "Culture is the foundation of European unity; it binds us together when pursuing shared objectives and underpins Europeans' sense of belonging to a collective project". Culture is challenging to define; a simple definition stands for "the way of life, especially the general customs and beliefs, of a particular group of people at a particular time",[1] but tactically we understand much more than this definition when talking about culture. Culture creates a belonging feeling to a place, shared gastronomy, shared traditions, shared sights, shared heritage and many more shared feelings. However, it is not a static concept. Culture evolves with society, culture is alive

---

[1]https://dictionary.cambridge.org/es/

B. de-Miguel-Molina (✉) · M. Segarra-Oña
Management Department, Universitat Politècnica de València, Valencia, Spain
e-mail: bdemigu@omp.upv.es

© The Author(s) 2018
V. Santamarina-Campos, M. Segarra-Oña (eds.), *Drones and the Creative Industry*,
https://doi.org/10.1007/978-3-319-95261-1_3

37

and creativity is one of its main characteristics. Culture is a network that generates economic, social and cultural impacts and that has to be supported.

The European Commission develops different initiatives to support Europe's cultural and creative sectors,[2] according to the Article 167 of the Treaty of the Functioning of the EU, which regulates the EU's role in the culture area. This area depends on the Directorate-General for Education, Youth, Sport and Culture department. One of this European Commission's framework programmes is Creative Europe,[3] which is devoted to supporting Europe's cultural and creative sectors by strengthening their innovativeness and competitiveness by enhancing specific skills and experience, including adaptation to digital technologies, enhancing the international cooperation or by creating new opportunities and business models.

Creative industries face two main problems for transforming their creativity and innovative skills into economic benefits and revenues: their small size and difficult access to funding (European Commission 2011). Moreover, the technological development leads to a fast digitisation (Acker et al. 2015) that has been widely supported by society, changing the way consumers access products and services, which is mainly affecting creative industries like music, books, movies and audiovisual works, which need to rapidly adapt their business models and value creation in order to survive. The analysis of the services' demand will identify the current customers' needs and help to orient their service and product offer. Moreover, and according to the European Creative Industries Alliance (2014), it is needed for someone to map and measure the effects and value added created by creative transfers to the wider economy. We will contribute to disentangling the creative industries' structure, how each subsector adds value and how the demand for services directly related to this industry is increasing, providing opportunities for establishing new businesses and taking advantage of the significant amount of knowledge that European cultural and creative industries generate.

2   The Market for Creative Industries

The wealth created by creative and cultural industries (CCIs) in the global economy reached 3% of the world GDP and generated 29.5 million jobs in 2013 (EY 2015). Concerning Europe, the CCIs also represented 3% of its GDP, and the associated employment accounted for 7.7 million in the same year.

In this section, each of the CCIs in the European market is explained in greater detail. To introduce them, Fig. 1 shows the weight of each sector in relation to the total value of the creative industries. The two variables included in the figure are the share of each sector in relation to the total employment and the share of each sector in relation to total revenues. The figure also shows that some sectors are more labour-

---

[2]https://ec.europa.eu/culture/

[3]https://ec.europa.eu/programmes/creative-europe/

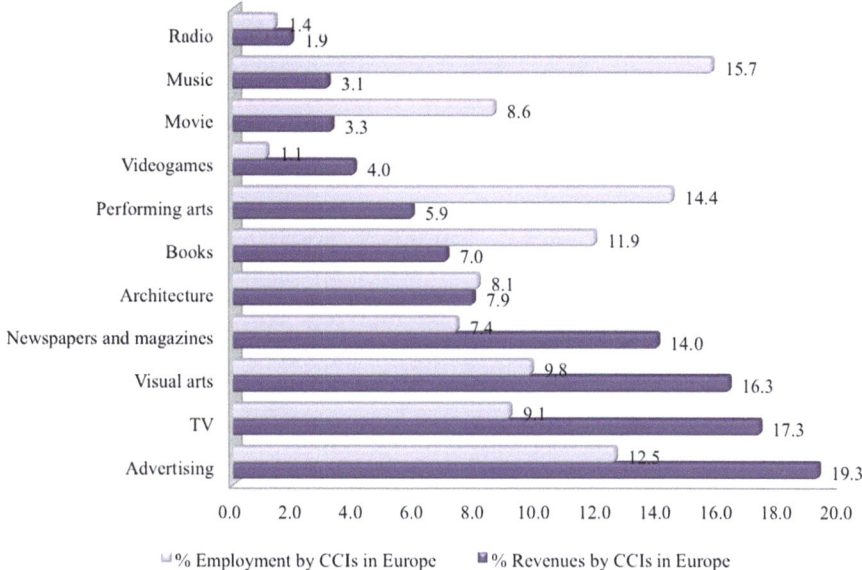

Fig. 1 Weight of each sector in the total value of CCIs in Europe. Source: Own elaboration from data in EY (2015)

intensive than others, as can be observed in music, performing arts and book sectors. In contrast, advertising, TV, visual arts and newspapers and magazines account for 67% of total revenues.

Considering the publishing industry as a whole, which includes newspapers, magazines and books, the percentage of revenues is similar to the percentage of employment of the total considered (27.3% employment and 21% in revenues), which positions the publishing industry as the most important within the creative and cultural industries.

2.1 The Publishing Market in Europe: Books, Newspapers and Magazines

The publishing industry, which includes books, newspapers and magazines, has been one of the industries most affected by the digitalisation phenomena (Acker et al. 2015). Despite this, the importance of the publishing sector in Europe remains, despite the effects of the economic crisis, as can be seen in Table 1. More precisely, and according to the Federation of European Publishers (FEP 2017), the book sector shows a recovery from 2014, while most of the growth of the business is due to the digital side (Acker et al. 2015). Among the data explaining the importance of this industry, it can be noted that (FEP 2017, Statista, Eurostat):

Table 1 Main indicators evolution: The publishing industry in the European Union

European Union (28 countries)	Y2012	Y2013	Y2014	Y2015	Var. Y2015/ Y2014 (%)	Main countries
N° Enterprises-Publishing of newspapers	7374	7654	7447	8120	9.0	France, Germany
N° Enterprises-Publishing of journals and periodicals	21,332	20,795	21,235	20,885	−1.6	France, UK, Germany, Italy
Employees: Publishing of newspapers	295,600	277,200	259,989	244,044	−6.1	Germany, UK, France
Employees: Publishing of journals and periodicals	200,900	186,600	189,364	180,633	−4.6	Germany, UK, France
Turnover or gross premiums written (million €)	38,295.6	36,743.7	36,773.3	37,754.6	2.7	Germany, UK, France
Turnover or gross premiums written (million €)	34,595.3	33,448	32,580.9	32,595.5	0.0	UK, Germany, France
Value added at factor cost (million €)	14,849.4	14,477.4	14,542.5	14,496.6	−0.3	Germany, UK, France
Value added at factor cost (million €)	15,029	14,255.1	13,887.1	13,368.9	−3.7	UK, Germany, France

Source: Eurostat, Structural Business Statistics

- Total annual sales revenue of book publishers in the EU (European Union) and EEA (European Economic Area) in 2015 were 22,300 million € and estimations for 2016 are that this value will increase to 22,500 million €.
- Concerning newspapers, although print circulation increased globally between 2012 and 2016, there was a decline of around −20% in Europe during this period (WAN-IFRA 2017).
- People employed in book publishing are 125,000, whereas the hole value chain employs around 500,000 people.
- The number of enterprises in the sector was 25,797 in 2008, while the number increased to 29,123 in 2014.
- The majority of the top publishing groups in the world are from Europe (6 to 8 of the top 10). Among the biggest European groups are Pearson (UK), RELX Group (UK/NL/US), Bertelsmann (Germany), Wolters Kluwer (NL), Hachette Livre

(France) and Grupo Planeta (Spain). In addition to large groups, small independent companies are common in the sector.

– The world's major book fairs take place in Europe (Frankfurt, London, Bologna).
– More than 500,000 new books are published by European groups every year. Concretely, the number of new books in 2016 was 590,000.
– The structure of revenues by category has been stable during the last years. These categories are four: (a) the trade/customer segment (48.4% of the net turnover), (b) school books (19.9%), (c) academic/professional (19.5%), and (d) children books (12.2%).

On the other hand, the sector has encountered different challenges during the last ten years, the majority as consequence of the economic crisis but also due to changes in technologies and their impact on the means available to read books. All these challenges have resulted in:

– Lower sales and revenues, due to the economic crisis. As a result, total annual sales revenue in the EU and EEA which had reached 24,500 million € in 2007, went down to 22,000 million € in 2014 when sales hit rock bottom. Reports (PwC 2016) state that the tendency in the sector is plane.
– People employed in book publishing were 169,400 in 2008 and 150,791 in 2014.
– More options for reading books, such as on mobiles and tablets, as e-book market emerged in 2007. Although the impact on printed books has not been as high as was predicted, the percentage of the e-book market ranges from 11% in the United Kingdom to 3% in Spain, with the average of the European market of 6%.
– Growth in self-publishing, books usually with lower prices and quality (PwC 2016).
– In some countries, the education authorities have promoted the use of books in digital format (PwC 2016).

In the last several years, publishers have needed to look for innovative solutions (Sandler 2017), including supporting start-ups to obtain ideas even from outside the book industry (Voigt and de Bruijn 2017). Some of the innovative projects include e-book signing technology and adding additional digital content to the books. In these solutions, digitalisation and other technologies are seen as opportunities instead of as threats.

2.2   The Advertising Market in Europe

The advertising industry is the second largest in terms of revenues (see Fig. 1) within the creative and cultural industries in Europe. The evolution of the sector is good, as all the analysed variables had a positive variation. The total number of companies belonging to the advertising industry surpassed 270,000 in 2015 with a workforce which exceeds 780,000 people. However, what seems more important when looking at the data (see Table 2) is that the value added has increased by 18 points from 2012 to 2015.

Table 2 Main indicators evolution: the advertising industry at the European Union

European Union (28 countries)	Y2012	Y2013	Y2014	Y2015	Var. Y2015/ Y2014 (%)	Main countries
N° Enterprises	243,564	248,658	263,866	270,548	2.53	Germany, Netherlands, Spain, Poland, France
Employees	770,000	776,400	747,869	780,496	4.36	Germany, UK, France, Spain
Turnover or gross premiums written (million €)	140,000	141,105.2	144,000	157,641.4	9.47	UK, Germany, France, Spain
Value added at factor cost (million €)	42,800	44,937.5	46,000	54,271.3	17.98	UK, Germany, France

Source: Eurostat, Structural Business Statistics

The advertising industry is usually split into different subsectors, such as Television advertising, Newspapers advertising, Magazine advertising, Cinema advertising, Radio advertising and Outdoor advertising, which confirm the traditional media. According to Statista (2018a, b), in 2016 the expenditure on traditional advertising in Europe (in million €) was as follows:

- 31,415 on Television advertising (↑).
- 14,535 on Newspapers advertising (↓).
- 6682 on Magazines advertising (↓).
- 737 on Cinema advertising (↑).
- 5148 on Radio advertising (↑).
- 5941 on Outdoor advertising (↔).

The digital revolution has also affected this sector and, since approximately 2006, there has been an explosion in digital advertising. Digital advertising includes online and internet advertising. The first one has increased from 6600 million € in 2006 to 41,900 million € in 2016.[4] Something similar occurred with internet advertising, which tripled its expenditure in the EU, from 13,586 million € in 2009 to 36,836 million € in 2016. The data show that digital advertisement is leading the industry and will surely define the trends and technological development that will affect the whole industry. This is why it is the main subsector to look at when looking for the identification of future needs.

---

[4]https://www.statista.com/statistics/436045/online-advertising-spending-by-format-europe/

Table 3 Main indicators evolution: the Radio and TV broadcasting industry in the European Union

European Union (28 countries)	Y2012	Y2013	Y2014	Y2015	Var. Y2015/ Y2014 (%)	Main countries
Radio						
N° Enterprises	6570	6557	6396	6243	−2.4	Spain, UK, Italy, Greece
Employees		58,900	58,774	55,734	−5.2	Germany, France, Spain, UK
Turnover or gross premiums written (million €)	8160.2	7951.5	8416.7	6771.7	−19.5	UK, France, Germany, Spain
Value added at factor cost (million €)	4575.3	4374.8	4437.6	4744.9	6.9	Germany, UK, France
TV						
N° Enterprises	5212	5159		5416	5.0	UK, Italy, Spain
Employees	183,400	189,500	177,248	182,350	2.9	Germany, France, Italy, Spain
Turnover or gross premiums written (million €)	58,354.2	57,602.2	59,580.8	62,603.7	5.1	UK, France, Italy, Germany
Value added at factor cost (million €)	22,521.6	22,522.,3	23,305.7	32,305	38.6	UK, Germany, France, Italy

Source: Eurostat, Structural Business Statistics

2.3   The Markets for Film, TV and Radio in Europe

In third place, with a 17.3% of revenues of the total CCI industries, we find the TV sector (see Fig. 1). This figure is close to 20% if we analyse TV and radio activities together.

In Table 3 we have displayed the main indicators for TV and radio. Data show that the value added of the TV activities expenditure is important when broadcasting is delivered on TV; however, radio is only getting back 6% of the total investment, while the rest of indicators are also decreasing, which does not suggest a promising future for the radio sector unless they introduce changes. These changes should be related to the digital revolution, including for example, internet-based radio, multimedia and on-demand audio. At the moment, the different technological options still coexist (DAB, DRM, DBM, etc.[5]) and there is no dominant model yet (Ala-Fossi et al. 2008; Delaere and Ballon 2017).

2.4   The Music Market in Europe

Although the critical situation that the music industry suffered in the late 1990s due to the technological and digital revolution (Dolata 2011) and related problems associated with copyright regulations (Dobusch and Schüßler 2014), the expected music industry revenue worldwide is increasing, moving from 37 billion € in 2012 to an expected 46.4 million € in 2021 (Statista 2018c).

Despite the continued decline in music sales in physical format, total revenues in 2016 grew by 6%, similar to the turnover increase in the European market. Digital incomes account for 50% of global music revenues and there are 112 million users of paid streaming subscriptions (IFPI 2017).

The leading markets in Europe for the music industry are the United Kingdom, Germany and France (see Table 4). However, when recorded music revenues per capita are considered, the most important countries are Norway, the United Kingdom, Sweden, Denmark and Germany (Statista 2018c). This data indicates that users in Nordic countries spend more on music than do users in other countries. For example, during 2014, Norwegian users spent 19.15 € in music, whereas French or Spanish users spent an average of 3.1 € on music (Statista 2018c).

The music industry is split into three parts: recorded, digital and streaming music. According to Statista (2018c), the evolution of each of them, in million €, is as follows:

- Recorded music, from 6600 in 2006 to 4300 in 2016 ($\downarrow$).
- Digital music, aggregated data not available ($\uparrow$).
- Streaming music, aggregated data not available ($\uparrow$).

The adoption of streaming formats in Europe is different depending on the country. For example, in Sweden (country of origin of Spotify) the revenues from streaming are 69% of the total, whereas in Germany physical sales are 52% of the total market (IFPI 2017). According to the Eurobarometer (2016), almost 71% of the respondents indicated that the main reason for selecting an online music service was that it was a free service, followed by good quality (53%).

The type of access for online music does not vary too much depending on the user's age and the video or music sharing websites with a variety of music/videos that are uploaded by individual users, artists or companies is the most used type of access. What is different depending on the user's age is the time spent listening to

---

[5]DAB, digital audio broadcasting; DRM, Digital Radio Mondiale; DMB, Digital Multimedia Broadcasting.

Table 4 Main indicators evolution: the sound recording and music publishing activities industry in the European Union

European Union (28 countries)	Y2012	Y2013	Y2014	Y2015	Var. Y2015/ Y2014 (%)	Main countries
N° Enterprises	23,080	25,000	25,820	25,614	−0.80	France, Sweden, UK, Netherlands, Germany
Employees	25,900	26,500		29,000	9.43	UK, Germany, France
Turnover or gross premiums written (million €)	7000	8001.5	8745	9100	4.06	UK, Germany, France
Value added at factor cost (million €)	3300	3911.9	4189.9	4400	5.01	Germany, UK

Source: Eurostat, Structural Business Statistics

digital music. In this case, people aged 15–24 spend almost two hours per day, while those in the 25–44 range, listen for approximately only one hour (Statista 2018c).

What we can extract from this analysis is that young people aged 15–24 years, who are the potential users in the near future, are totally adapted to the new music formats, digital and streaming, while traditional formats are rapidly declining.

3 Creative Industries and Emerging Technologies

During the last 15 years, internet and the digital technologies have impacted the cultural and creative industries, with special emphasis on publishing, TV and radio and music sectors. Although adapting to this change and constantly innovating, especially offering the final user new services options is necessary, it is most important to detect what the emerging technologies are that will impact the creative and cultural industries.

For the last two years, the biggest consulting firms and other important organisations have elaborated documents about emerging technologies and their use by creative industries. In this section, we have developed a content analysis of 27 documents elaborated by those firms and organisations. The documents are from the following sources and were published in the years 2016, 2017 and 2018:

– Accenture: 1 document (Accenture 2018).
– Arthur D. Little: 2 documents (Arthur D. Little 2017, 2018).
– Deloitte: 1 document (Deloitte 2018).
– EY: 2 documents (EY 2017, 2018).

- KPMG: 1 document (KPMG 2017).
- MIT Technology Review: 1 document (MIT Tech Review 2018).
- NESTA: 8 documents (NESTA 2017a, b, c, d, e, f, g, h).
- NESTA & Golant Media Ventures: 1 document (NESTA and Golant 2017).
- World Economic Forum: 7 documents (Harding 2016; Takahashi 2017; Hall and Takahashi 2017a, b; Hall 2018a, b; World Economic Forum 2018).
- World Economic Forum & McKinsey: 1 document (World Economic Forum 2018; McKinsey 2018).
- World Economic Forum & Accenture: 2 documents (World Economic Forum 2018; Accenture 2018).

To analyse information in the 27 documents, QDAMiner 4 software was used and the content analysed was developed (Provalis Research, Canada). Each document was read line-by-line to define the categories (15) and codes (160). The content analysis aimed to answer the following questions:

(a) What are the emerging technologies included in the documents and which creative industries are they for?
(b) Why do documents indicate that emerging technologies are important for creative industries?
(c) What challenges do firms in creative industries need to face in relation to emerging technologies?

Results obtained from the content analysis are included in the following subsections.

3.1 What Emerging Technologies and Creative Industries Are Included in the Documents?

The documents analysed different technologies, which have been organised into eight codes numbered form A1 to A8 (Fig. 2). Moreover, 21 codes were defined to indicate the creative industries cited in these documents and to indicate examples of current or potential applications of emerging technologies. Figure 2 presents the co-occurrences of codes for technologies and creative industries. Music and film industries were considered in more documents linked to artificial intelligence and digitalisation. Moreover, the newspaper industry was also considered in more documents about digitalisation, while music was included in more examples with virtual reality.

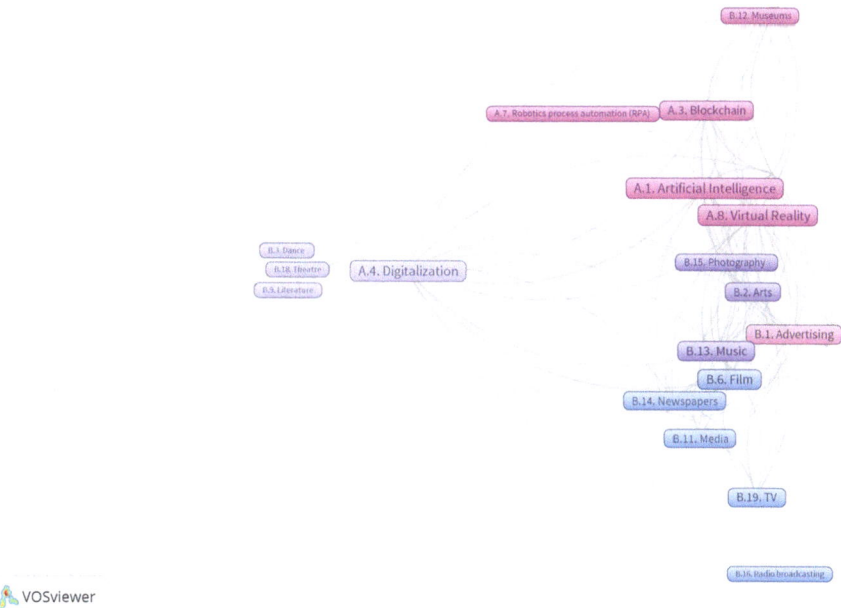

Fig. 2 Technologies by creative industries included in the documents analysed. Source: Own elaboration from content analysed of the 27 documents

3.2 Why Do Documents Indicate That Emerging Technologies Are Important for Creative Industries?

Through the content analysis, all the advantages and advances of emerging technologies were used to define different codes. These codes were grouped into four categories, one for each technology, as we explain in the next paragraphs.

Category C. Importance of Emerging Technologies for CIs in a General Sense This category includes information in the documents which did not refer to a specific technology, but to emerging technologies in general in this category; seven codes were defined that refer to what creative industries might obtain by incorporating some of these technologies in their business models, marketing and operations. The most cited code is the opportunity that these technologies offer for creating new experiences for customers. These codes, defined through the content analysis, are:

C.1. Distribute content to a broader audience (World Economic Forum 2018; McKinsey 2018).
C.2. Experiences in which digital and physical customers will converge (Arthur D Little 2017).
C.3. Increasingly integrated into the platforms where content is consumed (Hall and Takahashi 2017b).
C.4. New avenues for creativity (World Economic Forum 2018; McKinsey 2018).
C.5. New experiences for consumers (KPMG 2017; EY 2018; Accenture 2018; World Economic Forum 2018; McKinsey 2018).
C.6. Personalised content (Accenture 2018; World Economic Forum 2018; McKinsey 2018).

Category D. Artificial Intelligence (AI): Advantages and Advances In this category, we present the main ideas that the documents indicate about how artificial intelligence has been incorporated into different creative industries. Among the examples that documents give about AI application in creative industries are Recent.io (Hall and Takahashi 2017a), WordSmith (Hall 2018b), and Jukedeck (Harding 2016). Codes referring to advantages include opinions found in documents which indicated what can be obtained when AI is used, while advances refer to specific achievements in different creative industries. The main advantages found in the documents are included in the following codes, and results indicate that the main advantage is the use of users' preferences to create tailored content:

D.1. Analyse considerable datasets to learn specific behaviours (World Economic Forum 2018; McKinsey 2018).
D.2. Cloud-Based AI is making technology cheaper and easier to use (World Economic Forum 2018; McKinsey 2018).
D.3. Create tailored content through users' preferences (Hall and Takahashi 2017a; Hall 2018a; World Economic Forum 2018; McKinsey 2018).
D.4. Faster access to data and visualisations updated in real time (Hall 2018b).
D.5. General Adversarial Networks, GANs (MIT Technology Review 2018).
D.6. Many creative activities can be automated by AI (Hall 2018b).
D.7. Perform a task that is too difficult or time consuming for humans (Hall 2018a; World Economic Forum 2018; McKinsey 2018).
D.8. Products and services are much more personalised (Hall and Takahashi 2017a).
D.9. AI reduces the human element in the content creation process (Hall 2018b).
D.10. Turn structured data into a compelling text (Hall 2018b).
D.11. Understand how customers feel about products (World Economic Forum 2018; McKinsey 2018).

The examples for concrete creative industries show the advances obtained by firms in these sectors when they have applied AI. Among these examples, there are the following codes:

D.12. In the advertisement industry, detect fraudulent ad impressions (World Economic Forum 2018; McKinsey 2018).

D.13. In books, create novels (Hall 2018a).

D.14. In fashion design, generate new designs (World Economic Forum 2018; McKinsey 2018).

D.15. In films, write scripts and stage instructions (World Economic Forum 2018; McKinsey 2018; Hall 2018a).

D.16. In journalism, generate texts, saving journalists' time (KPMG 2017; World Economic Forum 2018; McKinsey 2018; Hall 2018b).

D.17. In music, compose music and produce instrumental sounds that humans had never heard before (Harding 2016; Hall 2018a; World Economic Forum 2018; McKinsey 2018).

D.18. In music, enable small-scale creators to use high-quality music at low cost (World Economic Forum 2018; McKinsey 2018).

D.19. In photography, generate sophisticated images (World Economic Forum 2018; McKinsey 2018).

D.20. In the media, the personalisation of news (Hall and Takahashi 2017a).

Category F. Augmented Reality (AR) and Virtual Reality (VR): Advantages and Advances In this category, benefits from using immersive technologies and their applications to creative industries are explained. Codes are defined from the analysis of content included in documents studied. These documents cite some examples of companies that have used immersive technologies. Some of these examples are Oculus VR (World Economic Forum 2018; Accenture 2018), Sony (Hall and Takahashi 2017b), Google (World Economic Forum 2018; McKinsey 2018), Facebook Spaces (Hall and Takahashi 2017a), Pokémon Go (Arthur D Little 2017), The Baltimore Ravens (Accenture 2018), The Denver Museum of Nature and Science (Accenture 2018) and The New York Times (Arthur D Little 2017).

Different advantages have been cited about firms in creative industries when using AR and VR, such as the option of creating immersive experiences through a smartphone. The information obtained and included in the codes about these immersive technologies is:

F.1. They are changing the way people connect with information, experiences, and each other (Accenture 2018).

F.2. They are tools for empathy and cognitive enhancement (Hall and Takahashi 2017a, b; World Economic Forum 2018; McKinsey 2018).

F.3. They can make content more powerful than when presented through traditional media (World Economic Forum 2018; McKinsey 2018).

F.4. Creators can replace rectilinear screens with full 360-degree fields of view (Hall and Takahashi 2017b; World Economic Forum 2018; McKinsey 2018).

F.5. Digital experiences using a smartphone (Arthur D Little 2017; Hall and Takahashi 2017a; World Economic Forum 2018; McKinsey 2018).

F.6. Smaller firms might create high-quality content at lower cost (Hall and Takahashi 2017b).

F.7. Moving from observation to immersion (Hall and Takahashi 2017a, b; Arthur D Little 2017).

F.8. New modes of experiencing content (Hall and Takahashi 2017b; World Economic Forum 2018; McKinsey 2018).

F.9. Reducing production costs in creative industries (Hall and Takahashi 2017b).

Codes indicating the use of AR and VR in specific creative industries are the following, and are referred to sectors such as design, music and publishing:

F.10. In design, higher accuracy and quality in products (Hall and Takahashi 2017b).

F.11. In design, shorten time and cost of iteration in new product development (Hall and Takahashi 2017b).

F.12. In music, share experiences with fans, including those who lack financial means to see live music (Harding 2016).

F.13. In music, sell merchandise through VR experiences (Harding 2016).

F.14. In publishing, have publishers differentiate their content offerings (World Economic Forum 2018; McKinsey 2018).

Category H. Blockchain: Advantages and Advances In this category, codes obtained in the content analysis related to blockchain are included. Blockchain appeared among the ten emerging technologies of 2016 cited by the World Economic Forum (2018). The Forum listed different challenges and opportunities related to the use of blockchain, such as cost reduction when intermediaries are displaced in transactions. On the other hand, big tech firms (such as Google) might reduce some costs linked to the current systems, reaching higher levels of security and privacy for their users. Finally, they also consider that blockchain technology would incentivise higher ethical behaviour among the participants.

There are increasing examples of the potential benefits of using blockchains in the creative and cultural industries. Among these industries, music is considered as the sector which would be most affected by blockchain technology (O'Dair and Beaven 2017). Some of the advantages referred to the use of blockchain in creative industries are:

- Accuracy and availability of copyright data about songs and recordings. From this view, identifying ownership would be easier. Examples indicating this advantage can be found for digital works (McConaghy et al. 2017) and recorded song (O'Dair and Beaven 2017).
- For digital works, provenance could be obtained and information could arrive at the original registration by the creator (McConaghy et al. 2017).
- Facilitation of near-instant micropayment for royalties (O'Dair and Beaven 2017).
- Higher transparency about the value chain (O'Dair and Beaven 2017).

In the content analysis undertaken for this chapter, we have found examples of the use of this technology such as Ethereum (Arthur D Little 2018), PeerTracks, Ascribe.io, Streamium, and Brave (Takahashi 2017).

The codes obtained through the analysis of information in the documents and referring to how artists could benefit from using blockchain are:

H.1. Allocate revenue to contributors (Takahashi 2017; World Economic Forum 2018; McKinsey 2018).

H.2. Blockchain allows "micro-metering" or "micro-monetising" (Takahashi 2017).

H.3. Allows artists to programme their IPRs, revenues and royalties into smart contracts (Takahashi 2017; World Economic Forum 2018; McKinsey 2018).

H.4. Blockchain avoids intermediaries in transactions (Hall and Takahashi 2017a; Arthur D Little 2018; World Economic Forum 2018; McKinsey 2018).

H.5. Creative content securely shared (Takahashi 2017).

H.6. Higher transparency (Takahashi 2017; World Economic Forum 2018; McKinsey 2018).

H.7. Creative work more accessible (Harding 2016; World Economic Forum 2018; McKinsey 2018).

H.8. New business models based on blockchain (Arthur D Little 2018).

H.9. Ownership can be traced (Takahashi 2017).

H.10. Artists can be paid based on information about actual consumption (World Economic Forum 2018; McKinsey 2018).

H.11. Ability to conduct dynamic prices (Takahashi 2017; World Economic Forum 2018; McKinsey 2018).

Category J. Digitalisation: Advantages and Challenges This category includes the codes defined which express benefits for firms in creative industries for integrating digitalisation in their organisations. This technology has had a significant impact on cultural organisations, as can be observed through NESTA works. Other advantages cited in the documents analysed include their impact on business models, marketing, efficiency and competitiveness. The codes defined for this category are the following:

J.1. Create new experiences and content for audiences (NESTA 2017g, h; NESTA and Golant Media Ventures 2017; Arthur D Little 2017).

J.2. Digital technologies are important for business models (NESTA 2017a, b, c, d, e, g, h; KPMG 2017; NESTA and Golant Media Ventures 2017).

J.3. Digital technologies are important for creation (NESTA 2017e; NESTA and Golant Media Ventures 2017).

J.4. Digital technologies are important in the digitalising of a collection in museums, preservation and archiving (NESTA 2017a, b, d, g).

J.5. Digital technologies are important for marketing (Arthur D Little 2017; NESTA 2017a, c, e, f, g, h; NESTA and Golant Media Ventures 2017).

J.6. Gain data about audiences and their behaviour (NESTA 2017h; NESTA and Golant Media Ventures 2017).

J.7. Lower costs and higher efficiency (Arthur D Little 2017; KPMG 2017; NESTA and Golant Media Ventures 2017).

J.8. Redefine customers' relationship (Arthur D Little 2017; World Economic Forum 2018).

J.9. Strengthen competitiveness (Arthur D Little 2017).

Category L. Extended Reality (XR): Advantages and Advances This category involves an additional emerging technology, and information compiled is focused in the document elaborated by Accenture (2018). Information codified from this document is divided into the three following codes:

L.1. This technology relocates people in time and space (Accenture 2018).
L.2. This technology immerses the user through visuals, audio, olfactory, and haptic cues (Accenture 2018).
L.3. XR expresses data in 3D environments, closer to the way humans see and imagine scenarios (Accenture 2018).

3.3   What Challenges Do Firms in Creative Industries Face About Emerging Technologies?

In this section, the content analysed focuses on the challenges that firms and professionals in creative industries face in relation to the emerging technologies analysed in Sect. 3.2. Technologies for which challenges were found in the documents analysed are artificial intelligence, immersive technologies (AR and VR), the blockchain and digitalisation. In the next paragraphs, codes defined for each technology are presented and advice is given about the challenges firms in creative industries need to consider for the next years. These needs include talent, business models, marketing and technical limitations, among others.

Category E. Artificial Intelligence (AI): Challenges Results from the content analysis indicate that AI systems are challenging to implement, that talent is difficult to find and that governance matters to assure trust. For specific industries, like journalism, AI could force journalists into obsolescence (Hall and Takahashi 2017a). Coexistence of AI and humans generates controversy when advantages in this technology are evaluated. Codes E.1. to E.7. cover the content analysis in this category:

E.1. Lack of AI talent as the most significant barrier (EY 2017).
E.2. AI use is dominated by a few big companies, such as Amazon, Google, and Microsoft (MIT Technology Review 2018; World Economic Forum 2018; McKinsey 2018).
E.3. Governance about disinformation and misinformation is an issue (Hall and Takahashi 2017a; World Economic Forum 2018; McKinsey 2018; Hall 2018a, b; Accenture 2018).
E.4. AI is not able to explain its output (Hall 2018b; World Economic Forum 2018; McKinsey 2018).
E.5. Platforms influence creative content (Hall and Takahashi 2017a; Hall 2018a).
E.6. AI disrupting markets and changing value chains for creative content (Hall 2018b; World Economic Forum 2018; McKinsey 2018).

E.7. AI is expensive and difficult to implement (EY 2017; MIT Technology Review 2018).

Category G. Augmented Reality (AR) and Virtual Reality (VR): Challenges In these technologies, challenges obtained through content analysis are primarily related to personal data that these immersive technologies can capture, and how the companies might use these data. More concretely, the codes defined in this category are the following:

G.1. Immersive technologies capture more intimate personal data from users (Hall and Takahashi 2017a, b; World Economic Forum 2018; McKinsey 2018).
G.2. One potential barrier to rapid progress is the lack of talent (Hall and Takahashi 2017b).
G.3. Devices require high-spec stationary computers (World Economic Forum 2018; McKinsey 2018).
G.4. Technical challenges, such as devices' size and battery life (Hall and Takahashi 2017b).
G.5. Price of VR headsets (World Economic Forum 2018; McKinsey 2018).
G.6. They could replace mobile computing (Hall and Takahashi 2017b).
G.7. Potential overuse like what occurs with mobiles (World Economic Forum 2018; McKinsey 2018).

Category I. Blockchain: Challenges Challenges cited by authors about the application of blockchain to creative industries were related to costs, knowledge needed, the low level of development of this technology and governance issues. Concerning specific creative industries, the examples referred to the music industry, and codes defined indicated that some musicians might not be prepared to do the job that was traditionally done by levels and publishers (Hall and Takahashi 2017a; Takahashi 2017, World Economic Forum 2018; McKinsey 2018). Also, if these traditional agents develop the blockchain infrastructure, there would be little change from previous remuneration systems (Takahashi 2017). Concerning codes regarding general challenges from blockchain, they have been defined as follows:

I.1. This is the least advanced of all the technologies and requires further development (Takahashi 2017; Hall and Takahashi 2017a; Arthur D Little 2018; World Economic Forum 2018; McKinsey 2018).
I.2. Any business that could stand to benefit from an immutable database can be disrupted by blockchain (Accenture 2018).
I.3. Blockchain-ready artists remain a small minority (Takahashi 2017).
I.4. IP concerns regarding whether the creative content is stored directly on the blockchain (Takahashi 2017).
I.5. Governance issues will be critical (Takahashi 2017; World Economic Forum 2018; McKinsey 2018).
I.6. Blockchain requires knowledge of alphanumeric code and cryptography (World Economic Forum 2018; McKinsey 2018).
I.7. Many stakeholders are uncomfortable with the level of transparency (World Economic Forum 2018; McKinsey 2018).

I.8. Costs regarding resources and time would be too high for creative applications (World Economic Forum 2018; McKinsey 2018).

Category K. Digitalisation: Challenges Firms in creative industries face digitalisation technologies, which allowed others to disrupt the sectors. Codes obtained from the content analysis indicate that the majority of challenges are related to Management and Strategic issues. Taking the right decision at the right moment seems to be an essential aspect that companies in creative industries have to consider. However, scarcity of resources might be an important limit for some organisations, especially in the cultural sector. In the next codes, information for challenges are further specified:

K.1. Existing business models are disrupted (Arthur D. Little 2017; NESTA and Golant Media Ventures 2017; World Economic Forum 2018).
K.2. Customer behaviour is changing (Accenture 2018; Arthur D. Little 2017; World Economic Forum 2018).
K.3. Digital technologies are used in all points of contact of the customer journey (Arthur D. Little 2017).
K.4. The early-mover/fast-follower advantage matters (KPMG 2017).
K.5. Services that previously were exclusively physical are now becoming increasingly digital (Arthur D. Little 2017).
K.6. Indecision around business models can be fatal (KPMG 2017).
K.7. Ecosystems and partner management are crucial in innovation and its application (Arthur D. Little 2018).
K.8. Funding, lack of resources and time, among important barriers to the digital development (NESTA 2017b, c, d, e, f, g).
K.9. Digital talent is crucial (World Economic Forum 2018).

4 Conclusions

In this chapter, two analyses of creative industries have been developed. The first analysis covers the study of the markets for creative industries in the last five years. For this purpose, variables like the number of employees, enterprises, revenues and value added have been used. Data indicate that the importance of these sectors in the European economy continues, despite the economic crisis. Digitalisation in these industries has been more a part of the solutions than it has been the cause of the problems, as can be observed in the sector such as publishing. However, this technology has boosted new business models based on self-publishing, thereby increasing the offering in the market and forcing incumbent companies to use new technologies across the entire value chain. In the advertising sector, digitalisation has increased internet advertising. Technologies have also impacted TV and music industries, in these cases spreading the use of streaming around European countries.

These technologies have changed how users consume TV and music content after disrupters entered the market, facilitating the shift to streaming.

The second analysis in this chapter focuses on the advantages of using emerging technologies and includes examples in specific creative industries. The advantages were explained through the content analysis of 27 documents of the most important consulting firms in the world. The documents were analysed, line-by-line, to search all the information indicating advantages, examples and challenges for technologies like artificial intelligence, augmented reality, virtual reality, blockchain and digitalisation. The analysis shows how these emerging technologies allow users to create and publish content for consumers in a different way, creating new experiences that are more personalised for users, and eliminating intermediaries in the creative process. AI, for example, can help with a task which is highly time consuming, while AR and VR increase the options for immersion in experiences. However, limitations occur in the implementation of these new technologies due to lack of talent, cost of investments, governance and technical issues. All these challenges are also included in the analysis.

References

Accenture (2018) Accenture technology vision 2018. Available via ACCENTURE. https://www.accenture.com/t20180222T121502Z__w__/us-en/_acnmedia/Accenture/next-gen-7/tech-vision-2018/pdf/Accenture-TechVision-2018-Exec-Summary.pdf. Accessed 15 Mar 2018

Acker O, Gröne F, Lefort T, Kropiunigg L (2015) The digital future of creative Europe. The impact of digitization and the Internet on the creative industries in Europe. Strategy&PWC. Available via PWC. https://www.strategyand.pwc.com/global/home. Accessed 22 Feb 2018

Ala-Fossi M, Lax S, O'Neill B, Jauert P, Shaw H (2008) The future of radio is still digital—but which one? Expert perspectives and future scenarios for radio media in 2015. J Radio Audio Media 15(1):4–25

Arthur D Little (2017) The customer meeting of the future. Arthur D Little, Paris. http://www.adlittle.com/sites/default/files/viewpoints/adl_the_customer_meeting_of_the_future.pdf. Accessed 15 Mar 2018

Arthur D Little (2018) Executive Roundtable: Digitalization. Arthur D Little, Paris. http://www.adlittle.com//Digitalization. Accessed 15 Mar 2018

Delaere S, Ballon P (2017) Standards, innovation and business models: the case of digital radio. In: Hawkins R, Blind K, Page R (eds) Handbook of innovation and standards. Edward Elgar, Cheltenham, pp 321–352

Deloitte (2018) Technology, media and telecommunications predictions 2018. Available via DELOITTE. https://www2.deloitte.com/global/en/pages/technology-media-and-telecommunications/articles/tmt-predictions.html. Accessed 15 Mar 2018

Dobusch L, Schüßler E (2014) Copyright reform and business model innovation: regulatory propaganda at German music industry conferences. Technol Forecast Soc Chang 83:24–39

Dolata U (2011) The music industry and the internet: a decade of disruptive and uncontrolled sectoral change. SOI Discussion Paper 2011–02. Research contributions to organizational sociology and innovation studies. Stuttgarter Beiträge zur Organisations-und Innovationsforschung

Eurobarometer (2016) https://ec.europa.eu/culture/policy/cultural-creative-industries_en. Retrieved from April 2018

European Commission (2011) Regulation of the European Parliament and of the Council establishing a creative Europe Framework Programme. Impact Assessment. Brussels, 23.11.2011 SEC(2011) 1399 final

European Creative Industries Alliance (2014) A new policy agenda to maximise the innovative contributions of Europe's creative industries. Recommendations from the Policy Learning Platform. Available via ECIA. http://eciaplatform.eu. Accessed 20 Feb 2018

EY (2015) Cultural times | The first global map of cultural and creative industries. http://www.worldcreative.org/. Retrieved from April 2018

EY (2017) Putting artificial intelligence (AI) to work. Available via EY. http://www.ey.com/gl/en/issues/business-environment/ey-innovation-matters-putting-artificial-intelligence-to-work. Accessed 15 Mar 2018

EY (2018) Opportunities and threats 2018. Media and Entertainment. Available via EY. http://www.ey.com/Publication/vwLUAssets/ey-opportunities-threats-2018-media-entertainment-summary/$File/ey-opportunities-threats-2018-media-entertainment-summary.pdf. Accessed 15 Mar 2018

Federation of European Publishers (2017) The book sector in Europe: facts and figures. Available via FEP. https://www.fep-fee.eu/The-Federation-of-European-862. Accessed 18 Feb 2018

Hall S (2018a) It's time to tackle tech's growing influence on the creative economy. World Economic Forum. Available via WEF. https://www.weforum.org/agenda/2018/02/its-time-to-tackle-techs-growing-influence-on-the-creative-economy. Accessed 15 Mar 2018

Hall S (2018b) Can you tell if this was written by a robot? 7 challenges for AI in journalism. World Economic Forum. Available via WEF. https://www.weforum.org/agenda/2018/01/can-you-tell-if-this-article-was-written-by-a-robot-7-challenges-for-ai-in-journalism. Accessed 15 Mar 2018

Hall S, Takahashi R (2017a) This four technologies will shape the creative economy—for better or worse. World Economic Forum. Available via WEF. https://www.weforum.org/agenda/2017/05/these-four-technologies-will-shape-the-creative-economy/. Accessed 15 Mar 2018

Hall S, Takahashi R (2017b) Augmented and virtual reality: the promise and peril of immersive technologies. World Economic Forum. Available via WEF. https://www.weforum.org/agenda/2017/09/augmented-and-virtual-reality-will-change-how-we-create-and-consume-and-bring-new-risks. Accessed 15 Mar 2018

Harding C (2016) 3 tech trends that will overturn the music industry—again. World Economic Forum. Available via WEF. https://www.weforum.org/agenda/2016/06/3-tech-trends-that-will-overturn-the-music-industry-again. Accessed 15 Mar 2018

IFPI (2017) IFPI Global Music Report 2017. Available via IFPI. http://www.ifpi.org/downloads/GMR2017.pdf. Accessed 15 Mar 2018

KPMG (2017) A call to action. Disruptive technologies barometer: media sector. Available via KPMG. https://home.kpmg.com/xx/en/home/insights/2016/12/disruptive-technologies-barometer-media-sector.html. Accessed 15 Mar 2018

McConaghy M, McMullen G, Parry G, McConaghy T, Holtzman D (2017) Visibility and digital art: blockchain as an ownership layer on the Internet. Strateg Chang 26(5):461–470

McKinsey (2018) Rebuilding corporate reputations, by Sheila Bonini, David Court, and Alberto Marchi. Accessible at https://www.mckinsey.com/featured-insights/leadership/rebuilding-corporate-reputations

MIT Technology Review (2018) 10 Breakthrough technologies 2018. Available via MIT. https://www.technologyreview.com/lists/technologies/2018/. Accessed 15 Mar 2018

NESTA (2017a) Digital culture 2017. Combined arts. Available via NESTA. https://www.nesta.org.uk/publications/digital-culture-2017-combined-arts. Accessed 15 Mar 2018

NESTA (2017b) Digital culture 2017. Dance. Available via NESTA. https://www.nesta.org.uk/publications/digital-culture-2017-dance. Accessed 15 Mar 2018

NESTA (2017c) Digital culture 2017. Available via NESTA. https://www.nesta.org.uk/publications/digital-culture-2017-literature. Accessed 15 Mar 2018

NESTA (2017d) Digital culture 2017. Available via NESTA. https://www.nesta.org.uk/publications/digital-culture-2017-museums. Accessed 15 Mar 2018

NESTA (2017e) Digital culture 2017. Available via NESTA. https://www.nesta.org.uk/publica tions/digital-culture-2017-music. Accessed 15 Mar 2018

NESTA (2017f) Digital culture 2017. Available via NESTA. https://www.nesta.org.uk/publica tions/digital-culture-2017-theatre. Accessed 15 Mar 2018

NESTA (2017g) Digital culture 2017. Available via NESTA. https://www.nesta.org.uk/publica tions/digital-culture-2017-visual-arts. Accessed 15 Mar 2018

NESTA (2017h) Digital culture 2017. Available via NESTA. https://www.nesta.org.uk/publica tions/digital-culture-2017. Accessed 15 Mar 2018

NESTA & Golant Media Ventures (2017) The adoption of digital technology in the arts. Available via NESTA. https://www.nesta.org.uk/sites/default/files/difaw_gmv_e.pdf. Accessed 15 Mar 2018

O'Dair M, Beaven Z (2017) The networked record industry: how blockchain technology could transform the record industry. Strateg Chang 26(5):471–480

PwC (2016) Entertainment and media outlook 2016–2020, España. Available via PWC. https:// www.pwc.es/es/publicaciones. Accessed 15 Mar 2018

Sandler K (2017) Innovation in publishing: this is not an oxymoron! Publ Res Q 33:328–342

Statista (2018a) Advertising in Europe, New York. https://www.statista.com. Accessed 15 Mar 2018

Statista (2018b) Digital advertising in Europe, New York. https://www.statista.com. Accessed 15 Mar 2018

Statista (2018c) Music industry in Europe, New York. https://www.statista.com. Accessed 15 Mar 2018

Takahashi (2017) How can creative industries benefit from blockchain? World Economic Forum. Available via WEF. https://www.weforum.org/agenda/2017/07/how-can-creative-industries-benefit-from-blockchain. Accessed 15 Mar 2018

Voigt M, de Bruijn W (2017) Dutch publishing industry seeks startups through the renew the book competition. Publ Res Q 33:10–13

WAN-IFRA (2017) World Press trends 2017. Available via WPT. http://www.wptdatabase.org/world-press-trends-2017-facts-and-figures. Accessed 15 Mar 2018

World Economic Forum (2018) Disrupting business models is not enough. We need tech innovation too. Accessible via WEF at https://www.weforum.org/agenda/2018/03/sharing-economy-product-innovation-balance-disruption/. Accessed on May 2018

Financing Tech-Transfer and Innovation: An Application to the Creative Industries

Ana-Cruz García, Miguel Pizá, and Francisca Gómez

Abstract A key point for tech-transfer and innovation is financing, especially for small and medium enterprises. In this case, the authors explore the different European instruments of financing and how they have been applied to the creative industries. Starting from an analysis of the type of projects funded by previous European Framework Programmes for Research and Innovation and their evolution, the authors identify challenges and risks for the sector.

1 Introduction: Technology and Innovation in Creative Industries

The Creative Industries Sector (CIS) is among the most innovating sectors both in terms of technological and non-technological innovation. As for the rest of industries, technology plays a very important role in the innovation capacity of the sector. In terms of technological innovation creative industries (CI) have a twofold role, acting both as user and as provider of technologies. Moreover, CIS has an increasingly recognised role in triggering technological innovation in other sectors and in the society as a whole (Bakshi et al. 2008).

CI are strong users of new technologies. According to an Austrian CI survey (Müller et al. 2009), more than 90% of creative industries use new technologies in their daily business. Additionally, according to the same survey, 46% of CI companies that acquire new technologies make contact with technology producers before acquiring it. This direct contact supports innovation of the technological sector and supports the idea that for CI it is important to collaborate with technology providers in order to assimilate technologies. As a matter of fact, 18% of CI reported that the new technology was developed specifically for them, acting also a source of technological innovation.

A.-C. García (✉) · M. Pizá · F. Gómez
Fundación Ciudad Politècnica de la Innovación, Universitat Politècnica de València, Valencia, Spain

© The Author(s) 2018   59
V. Santamarina-Campos, M. Segarra-Oña (eds.), *Drones and the Creative Industry*,
https://doi.org/10.1007/978-3-319-95261-1_4

The most demanded technologies by CI are Information and Communication Technologies (ICT) both in terms of software and hardware, but also new materials in the case of design fashion and performing arts (Müller et al. 2009). As in other industries, digital transformation has had a profound impact in CI. In terms of product development, content created by CI is, in many cases, digital or digitised; however, CIs have also invented new business models that integrate high-value content and communication technologies, both within their business and the services they provide.

Some CI sectors are frontrunners in introducing consumers to the new business models of the digital era, paving the way for other sectors that are shifting services online. Together, they identify and fulfil new uses for technology and adapt to changing consumer behaviour and expectations. Finally, once seen as largely a matter of technology, innovation has evolved to embrace aesthetics, functionality and content, and CI can act as a provider of them especially in ICT technologies. According to the above-mentioned study (Müller et al. 2009), 17% of the creative enterprises provide support to Research and Development (R&D) and engineering design.

Moreover, creative industries may also support innovation in the wider economy without direct interaction. A key mechanism for doing so is the mobility of the workforce, in particular when people find new jobs outside the creative industries and take their ideas, knowledge and creative potential with them and use it in other industries. There are also significant Business to Business (B2B) linkages between creative industries and the wider economy and they show that firms with a higher share of inputs from creative industries tend to be more innovative in terms of product innovations (Bakshi et al. 2008; Experian 2007; Müller et al. 2009).

CI are also developers of technologies that can be applied in other sectors. A good example of this is digital gaming and animation, which is at the forefront of technology, and which has developed many technologies that have been transferred to other sectors inside and outside of CI. The enormous success of digital games in engaging users (McDonald 2017) and the possibilities that digital gaming offers in terms of information and learning has fostered the use of gaming in different areas, such as in learning and patient involvement. This fact has opened a new market for the digital gaming industry and has created a new subsector in the area of serious games.

Universities can also play an important role in this ecosystem of innovation and technology transfer from and to CI. As with any sector, universities have a double role regarding creative industries, namely training and education, but also as technology developers. Successful innovation in CI most often requires the combination of a firm's own innovation resources with external inputs. Universities can provide those inputs in the form of knowledge (e.g. developed technologies, insights from studies, etc.) or specialised R&D services. CI demand for new applications, especially in the area of ICT, can provide a major stimulus for innovation on the side of the technology producers, such as universities.

2   Challenges

Technology-based innovation in CIs is rather complex. Despite the already mentioned use of new technologies, within CIs there is a generalised lack of knowledge of technologies that could be applied for the benefit of creative industries both from the side of technology providers/research groups and from the side of CI. Although some technologies, such as Virtual Reality (VR) and Augmented Reality (AR) have been widely applied in many CI sectors, there is a wide diversity of technologies that could be applied to increase the competitiveness of CI that are not recognised as such. This is especially important when we talk about very small firms of small and medium enterprises (SMEs) and technologies that are not as "popular" as VR and AR.

Moreover, there is a range of already developed technologies that cannot be fully exploited for CI due to the lack of adaptation of their particular needs and workforce skills. In some cases, technology-based innovation in CI is more about adoption than about development (Universities UK 2010). This is typical for many SME sectors and there are examples, such as in manufacturing, where big efforts are being made to put the technology in the hands of the SMEs by delivering specialised tools and infrastructures. The particularities of CI make necessary for this adaptation to be tailored to the specificity of the different CI subsectors.

One of the reasons of this lack of knowledge and adaptation of technologies is the lack of sectoral collaboration between CI and other sectors and even the lack of structured R&D collaboration within the CI. If we compare the situation of CI in Europe with that of many other sectors such as manufacturing, construction or energy, the connection between researchers and industry is very scarce and much less structured. The exception is the New European Media Initiative's platform (NEM) that has recently assumed the CI sectors and created a working group about CI.

In relation to universities, in most cases, education and R&D structures, such as research groups, are focused either on technological areas, such as ICT, chemistry, physics, etc., or on application areas, such as industrial sectors or arts. The multidisciplinary approach and the combination of Arts and Technology as required by many CI has been traditionally scarce. Besides, in many cases, universities focus more on linear than on collaborative processes, which are required to work with CI (Universities UK 2010). This imposes difficulties for both the technology transfer and education/training role of Universities regarding CI.

3   European Funding of Research and Innovation for the Benefit of CIS

Framework Programmes (FP) for Research Technology Development have been, during the last few decades, the main source of Research and Development Collaboration among the academy and industry at the European level. Horizon 2020

(H2020) is the running Framework Programme for Research and Innovation, which was launched in 2014 by the European Commission. Horizon 2020 is structured under three pillars: Excellent Science (ES), Societal Challenges (SC) and Industrial Leadership (LEIT) together with other areas, such as Science with and for Society (SWAFS) and Spreading Excellence. For the first time, it combines research and innovation in a single programme under a holistic approach to innovation that can, at the same time, tackle societal challenges and give rise to new competitive businesses and industries.

Under this approach, innovation has a broad definition, which is not only limited to the development of new products and services on the basis of scientific and technological breakthroughs. Under the innovation approach of H2020 there is a particular emphasis on supporting activities which operate close to the end-users and the market, such as demonstration, piloting or proof-of-concept. According to this approach, different types of projects, called instruments, are funded by H2020. On one hand, the Research and Innovation Action (RIA) instrument covers the activities that explore the feasibility of a new technology process or product and the testing at a small scale in laboratory or simulated environment. On the other hand, Innovation Actions (IA) cover the activities of adaptation and demonstration in operational environment of a new, but already existing, technology for a given application. Complementary to this, another type of project called Coordination and Support Action (CSA) is directed at activities such as networking, awareness raising, mutual learning exercises and strategic planning.

It is important to mention that in Horizon 2020, as well as in the previous FP6 and FP7 programmes, the calls for proposals are quite restrictive. The calls include a number of topics that determine the technologies and applications that will be funded. This implies that, in some way, the support given to a technology and sector is predetermined by the description of the topics.

The support given to CI by the different European Framework Programmes, according to CORDIS database, shows a clear evolution (Fig. 1). FP5 (1998–2001) only funded one project, while in FP6 (2002–2007) only two projects were funded. The support to CI increased significantly in the period 2008–2013 under FP7, and the (CIP), which was a complementary Programme focused on innovation. In total 35 projects were funded in that period, 26 under FP6 and 8 under CIP. On the other hand, as can be seen in Fig. 1, the number of projects funded under H2020 (2014–2020) has increased exponentially. Even before finishing H2020, by 2017 100 projects were already funded.

If we analyse the area of H2020 in which the projects were funded (see Fig. 2) we can see that most of the projects correspond to calls of the Leadership in Enabling and Industrial Technologies (LEIT) and in particular in the area of ICT with a total of 85 projects funded by 2017. The diversity of ICT technologies in projects related to CI that were funded by H2020 is very broad, covering among others: Visualisation, Language Technologies, Multimodal and Natural Interaction, Big Data for Content Management, Content Creation and Annotation, Serious Games for Learning, Wearable technologies and the impact of Future Internet among others. The projects covered both Research and Innovation Actions (RIA) and Innovation Actions (IA).

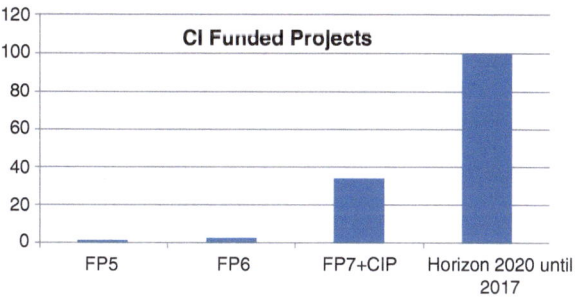

Fig. 1 Number of funded projects with contribution of creative industries in the different European Frame Work Programmes. Source: CORDIS under search term "creative industries"

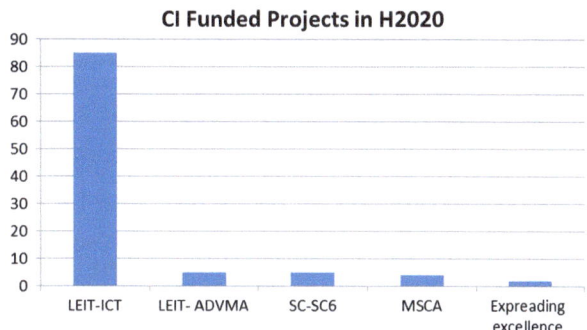

Fig. 2 Number of funded projects with contribution of creative industries in the different areas of H2020 until 2017. Source: CORDIS under search term "creative industries"

The other areas with funded projects include the one in LEIT about Advanced Materials (ADVMA) and the Societal Challenge 6 "*Europe in a changing world-Inclusive, innovative and reflective societies*", where most of the research about cultural heritage and cultural industries is included. Both areas funded five projects each, before 2017. This distribution of funded projects on ICT and materials technologies corresponds to the already mentioned results of Müller et al. (2009) regarding the impact of technologies in CI, and also reflects the support given to CI for the different areas of H2020 in terms of the scope of the launched calls for proposals.

It is also worth mentioning that only four projects in total were funded under the Marie Slodowska Curie Actions (MSCA). This MSCA scheme is devoted to improving mobility of researchers. Of those four MSCA CI projects funded, two of them were RISE, which support the interchange of personnel among academy and industry, and two of them were individual fellowships. No project was funded under the International Training Network (ITN) scheme whose objective is to raise excellence and structure research and doctoral training in Europe, extending the traditional academic research training setting. ITN scheme is open to any area of research but, as data about submitted proposals are not available, it is impossible to know if no proposals were submitted or if the proposals submitted did not succeed in receiving funds.

In 2014, 2015, 2016 and 2017, specific calls were launched for the support of the creative industries inside of the LEIT-ICT area of Horizon 2020. This specific

Fig. 3 Number of projects funded in the period 2014–2017 in specific calls of LEIT-ICT Horizon 2020 for CI. Source: CORDIS under search term "creative industries"

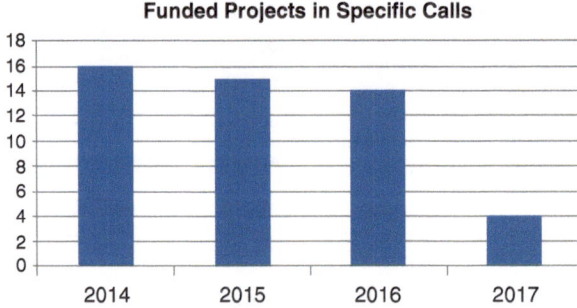

Fig. 4 Number of projects funded by type of action in the period 2014–2017 in specific calls of LEIT-ICT Horizon 2020 for CI. Source: CORDIS under search term "creative industries"

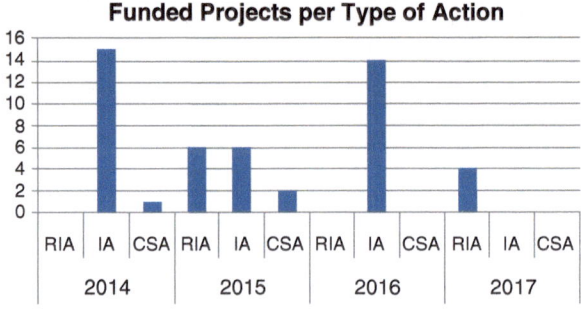

activity funded 49 projects in total, including RIA and IA actions, according to the budget allocated to each type of instrument. In Figs. 3 and 4, the number and type of projects funded per year are presented.

As can be seen in Fig. 3, the number of funded projects has decreased over the years. This is due to a combination of a reduction of budget and the mean budget per project (see Figs. 5 and 6). As can be seen in Fig. 4, in 2014 and 2016, the number of funded projects was quite high, even when the budget was similar to that of 2017 (Fig. 5). This is due to the fact that the calls of these years included topics for small Innovation Projects with low budgets of around 1 million euros. These topics were specially designed to increase the competitiveness of CI by launching to the market products and services based on technologies that have reached a level of demonstration in a relevant environment (Technology Readiness Level 6). The project itself should adapt the technology to the special needs of creative industries and develop the necessary industrialisation and demonstration to surpass the so-called "Valley of Death". This term is a metaphor that illustrates the challenges faced by a company (normally a spin-off company) to go from a functional prototype to a fully-fledged profitable business. This term is used mainly in the case of spin-off companies, but can also be applied to products and services developed by any SME and can be perfectly applied to CI SMEs.

In Horizon 2020, this concept of "Valley of Death" was at the core of the creation of the Programme as focusing on Research and Innovation. As a matter of fact, some specific actions were designed to help SMEs to surpass this "Valley of Death". These

Fig. 5 Budget specific calls of LEIT-ICT Horizon 2020 for CI in the period 2014–2017. Source: CORDIS under search term "creative industries"

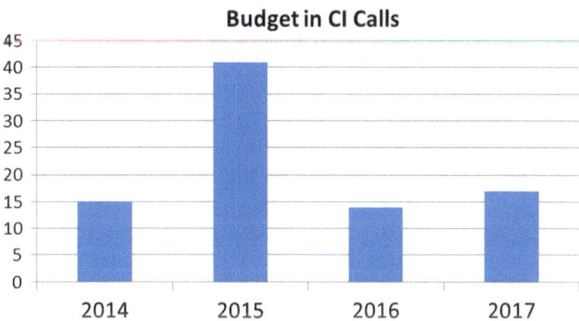

Fig. 6 Mean budget of funded project per year in CI specific calls of LEIT-ICT Horizon 2020. Source: CORDIS under search term "creative industries"

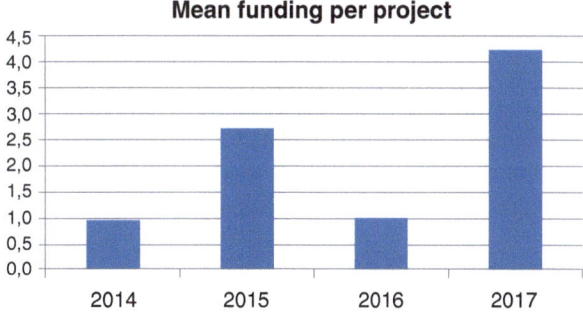

specific instruments are the so-called SME Instruments, and in some way they are also the Fast Track to Innovation. These types of projects are Innovation Actions conceived of as bottom-up actions in the sense that the calls are open to very broad objectives and technologies.

Although SME Instrument and Fast Track to Innovation could, in principle, have been a good opportunity for CI, in the period from 2014 to 2017, only three CI projects among 3200 were funded under this scheme. Besides, no IC project was funded under the Fast Track to Innovation Scheme. As the information about the submitted proposals is not available, we cannot know if the reason for this surprising result is that few proposals were submitted or if these proposals were not able to compete with those from other sectors.

In this context, the specific support of H2020 funding small projects very focused in the transfer of technology to the CI is remarkable. In particular, the strong support of the ICT area of Horizon 2020 has to be recognised. In 2012, a "Creativity" unit was set up in 2012 following a reorganisation of Directorate-General CONNECT Communications Networks, Content and Technology (DG CONNECT). The unit was part of Directorate G: Media & Data. Until 2016, this Creativity Unit supported ICT-related research and innovation for CI and research on creative processes involving ICT.

Nevertheless, in 2016 a reorganisation of the DG CONNECT led the Creativity Unit to be transformed into a new one that deals with a broader area for Data Applications and Creativity. Additionally, the specific calls for CI have disappeared from calls for proposals 2018–2020. For this final period of H2020, the calls in which CI are mentioned are not specific to them and CI have to compete with other sectors that have a much longer tradition of securing R&D EU funding, such as manufacturing and health. These other sectors are also much more structured and involve big companies that have been Key Players of European Research for decades. This scenario implies the risk that, if the special needs of CI are not taken into account, CI sectors will lag behind in terms of technological innovation and technology providers will lose CI as a potential market.

4   Discussion

In terms of research and development, many challenges have to be overcome in order to foster technological innovation in CI. There is a lack of awareness in academia about the possibilities for the CI as potential users of a great diversity of technologies, and in particular, digital technologies. Moreover, there is also a lack of knowledge in many CI about the technological possibilities that can be applied to the sector, especially in the case of small SMEs. As a consequence, there are no clear roadmaps of R&D required for CI that can guide the creation of R&D roadmaps in research groups or funding organisations.

To be effective, appropriate funding for CIs' technological innovation should be reinforced by mechanisms that are able to reach to such a disperse sector. This has deep implications for innovation policy: no longer is it sufficient to support the creative industries alone and for their own sake, policy should encourage and embed linkages between them and the wider economy.

But there are not only R&D challenges to face. Skills are maybe the most crucial input to industrial innovation, but skilled and talented people are a key element for a firm's potential to absorb external knowledge. A crucial part of the innovation-stimulating and supporting potential of the creative industries certainly lies within the human capital of their workforce. Nowadays, a strong demand in terms of education and R&D has arisen from the growth of the CI sector. This has led to many private and public universities developing education offers in multidisciplinary grades and master's level courses that are focused in covering needs of CI. In addition, multidisciplinary research groups that have the potential to develop technologies and create contents specifically conceived for the use of CIs have recently appeared.

Nevertheless, although in some areas, such as in video gaming, several training programmes have been developed, there still exists a gap between the artistic and the technological training programmes at grade and master's levels, not to mention in doctorate degrees. To cover this gap, Mixed Art-Technology Programmes focused on the needs of CI should be developed, including at the doctoral level.

5 Conclusions

The support of Research and Development Programmes is key to boosting the competitiveness of any sector, but they are even more crucial for very important, but not so well structured, sectors such as CIS. From a European perspective, the recognition and support that has been given to the sector by the European Commission and in particular in the ICT area (DG CONNECT) has to be acknowledged. Nevertheless, there is still a lot to do and the new approach of the WP2018–2020 has the risk of once again leaving CI far behind other sectors in terms of support for the development of new technologies and products and the corresponding innovation activities. In addition, the low number of projects funded under bottom-up actions, such as SME Instrument, FTI or MSCA actions, should be carefully studied to know the reasons why. In this sense, actions should be taken to improve the participation of CI SMEs that want to incorporate new technologies in their innovation processes or products. Besides, actions should be taken to foster the development of innovative doctorate training programmes under the MSCA scheme for the benefit of CIS.

References

Bakshi H, McVittie E, Simmie J (2008) Creating innovation. Do the creative industries support innovation in the wider economy? NESTA Research report July 2008. NESTA, London

Experian (2007) How linked are the UK's creative industries to the wider economy? An input–output analysis, NESTA Working Paper, London

McDonald E (2017) The global games market will reach $108.9 billion in 2017 with mobile taking 42%. Available via NEWZOO. https://newzoo.com/insights/articles/the-global-games-market-will-reach-108-9-billion-in-2017-with-mobile-taking-42/. Accessed 15 Mar 2018

Müller K, Rammer C, Trüby J (2009) The role of creative industries in industrial innovation. Innovation 11(2):148–168

Universities UK (2010) Creating prosperity: the role of higher education in driving the UK's creative economy. Available via UNIVERSITIES UK. http://www.universitiesuk.ac.uk/policy-and-analysis/reports/Documents/2010/creating-prosperity-the-role-of-higher-education.pdf. Accessed 15 Mar 2018

Successful Cases of the Use of Innovative Tools and Technology in the Creative Industries Field

María-Ángeles Carabal-Montagud, Virginia Santamarina-Campos, Gavin O'Brien, and María de-Miguel-Molina

Abstract The aim of this chapter is to present successful examples of technology and the creative industries complementing each other and working together, as well as to highlight how the transfer of tech-knowledge can be applied to creative industries, such as filming. In order to do this, we analyse in detail successful case studies between new technology and innovation in general, discussing some examples that are either currently emerging or fully consolidated. Subsequently, we explore the successful use of drones within creative industries in recent years, across a variety of culturally important industries such as professional photography, aerial photography and filming, advertising, the film industry, television, performing arts, video gaming, architecture and heritage. The aim of this to highlight the impact they generate in the creative sector, which in turn has led to a revolution in the means of generating content and creative experiences.

1 Relationship Between Technology and Creative Industries

The CI is the sector that most uses new and innovative technologies and tools to make progress in its projects, in order to search for opportunities at a business level. If there is something that distinguishes different sectors within the creative industries from other industries, it is their entrepreneurial capacity and heightened use of innovative tools.

M.-Á. Carabal-Montagud (✉) · V. Santamarina-Campos
Conserv. & Restoration of Cult. Heritage Department, Universitat Politècnica de València, Valencia, Spain
e-mail: macamon@crbc.upv.es

G. O'Brien
Clearhead, Luton, England

M. de-Miguel-Molina
Management Department, Universitat Politècnica de València, Valencia, Spain

V. Santamarina-Campos, M. Segarra-Oña (eds.), *Drones and the Creative Industry*,
https://doi.org/10.1007/978-3-319-95261-1_5

It could be said that they serve as a link between technological resources and the cultural fabric, of society. The CIs have the possibility of generating creative discourses, which reach a wide sector of society and are echoed by social networks. These industries spark interest across these channels due to their capacity to be thought-provokingly innovative and, as such, technology and innovation and the creative industries are able to share the interest of the general public. Since they have been united in recent years, they have managed to change the context and the content they represent, making it more playful and attractive, succeeding in appealing to an ever-greater majority of the public, motivating them and generating an expectation with respect to novelties, due to the speed with which advancements are made.

1.1 The IC: Social and Economic Context

Creative industries account for 3% of the world GDP, according to the International Confederation of Societies of Authors and Composers (CISCAC 2015), which implies an economic impact of almost 1835 billion euros, generating employment for 1% of the society, around 29.5 million people, according to the aforementioned report.

Within Europe, the same report shows that 7.7 million people currently work within creative industries, with about 578 billion euros in revenue, making it the second largest CCI market and number one in advertising (CISCAC 2015).

Taking these figures into account, the economic impact of this sector immediately redounds to the social fabric. This economic data clearly indicates very high revenue and shows that focus on culture and sectors which have great culturally symbolic value are a driving force for development that generates income both directly and indirectly.

The Global Report on Cultural Policies: Focus on UIS Data and Analysis (UNESCO 2017), points out that digital distribution platforms, exchange networks and export strategies, mainly in the audiovisual sector, are helping Southern countries to enter into the international market of cultural goods and services. As such, an economic impact linked to the development of the different regions is seen.

Snowball demonstrates that culture plays an essential role in many fields that are inherent to it, such as entertainment and education, as well as in creating and stimulating social awareness, without losing sight of its cohesive value to social collectives (Snowball 2016). Culture transcends any economic model, however, and when discussing creative industries, it is necessary to take into account business models that are economically viable and profitable. Culture is a universal basis for all societies and is capable of generating inclusive social development based on very fragmented realities. It has an impact on the entirety of society, on education and on the transmission of ideas, and sometimes results in less-quantifiable data, such as social enrichment.

After evaluating the impact of culture, CI is projected in society, having results which are highly valuable to society. This value motivates various governments to generate public economic aid to invest in their development, thus supporting production.

1.2 Use of Innovative Tools and Technology in the CI

According to the European Economic and Social Committee, CI must have a technological component and also add creative value that protects intellectual property rights in the internal market. Additionally, CI must promote the search for new technologies and innovative uses of products and processes on an international scale. These processes must meet the European quality regulations and therefore guarantee the development of value chains through networks and common distribution systems (Pezzini and Konstantinou 2013).

It is also important to point out that some of the common challenges of CI include an increasing impact on the production and distribution processes caused by the conversion to a digital system and the diffusion of new technologies, and the demand for greater synergies between the creative world and cultural and technological innovations (Pezzini and Konstantinou 2013).

According to Wang (2012), the creative industry "is an emerging industry formed on the basis of integrating and using information technology and relevant literature-based business models in combination with research analysis regarding various factors in the industry".

2 Successful Uses of Innovative Tools and Technology in the CI

In the following section, we analyse various successful cases of innovation tools in the creative industries and the impact of technologies in the various areas that make up these industries in general. Also in this section, we take a look at the infinite possibilities offered by technology regarding creative industries. Some instances of this occurring demonstrate the development, expansion and visibility of these technologies in the CI. These drones are one of the most important technological assets of recent years, considering the use of RPAS has disseminated in CI.

2.1 Successful Examples in CI in General

New technologies have revolutionised the creative industries, given that it is the field where they are most visible and where they are most applicable. It is not surprising that specific terminology has emerged that unites both fields as the so-called "digital artists".

We could list a multitude of success stories, such as the use of 360° videos and virtual augmented reality in the cinematographic or publicity field. There is also the additional example of curved television screens to provide an immersive experience, improving the panoramic effect and giving an experience closer to that of reality. Another noteworthy use of technology is audiovisual and interactive installations, experimental electronic music and multimedia presentations.

In these categories, we observe the growing incorporation of art and technology in companies destined to the movie industry, such as Pixar®. Art and technology comes together through interactive Video Mapping, used to create virtual spaces in 3D and 4D, for various mediums such as dance, sculpture, performances or reproductions of existing architecture, with the unification of illustration and new technologies or interactive Video Mapping creating virtual spaces in 3D and 4D for dance, sculpture, performance and the reproduction of architectural works. Video Mapping has multiple uses in CI (Higgs and Cunningham 2008), and it is also used in the field of filmmaking and videogames, in which it facilitates the creation of immersive virtual spaces.

The example of electronic textiles applied to fashion has also gained special relevance in recent years. Electric clothes and smart clothes have taken form as so-called "wearables" or smart devices applied to haute couture. An example of this is Adrenaline Dress®, which incorporates an Intel® chip that interacts with various moods. Another example is the creation of an intelligent shelter that incorporates infrared technology or fabrics that measure the pollution of the air. Google Jacquard ® has taken initiative in the field of the textile industry, creating the first garments that integrate the understanding of tactile gestures capable of activating digital services. As Google® indicates in the description of the clothing created with textile raw materials, together with technological materials, "By starting with raw materials, such as yarns and textiles, we found ways to provide unprecedented access to the digital world through items that are not typically considered to be technology. So, your most beloved items—a favourite jacket, a pair of shoes, the bag you take everywhere—will keep you connected to your digital life in new, seamless ways" (Google 2015). The idea arose in 2014 and, together with the company Levi's®, they have been working on this innovation, which was presented at the Cooper Hewitt Smithsonian Design Museum in 2016, for their permanent collection as a part of the history of American Design (Google 2015). Their technological garments can be purchased through their website.

Also from Google® came Magenta® projects, which uses neural networks for the creation of musical and artistic compositions, and Deep Dream® software that uses algorithms and adjusts them to the neural network to create psychedelic images.

Another function of this technology and a great success in innovation is the ability to bring the creative industries closer to spectators who have disabilities and allow them to enjoy them as well. 3D Printers have revealed themselves as one of the most capable technological tools to offer themselves in a versatile way in a myriad of fields, including breaking down sensory barriers. Examples of this are cases such as transferring works of art, photographs or comics, among other works, into a tactile format so that the blind can enjoy works that they otherwise would be able to experience. The Prado Museum in Madrid presented the Touching the Prado project in 2015, being the first initiative to reproduce works with 3D printers, aimed at visually impaired visitors, so that blind people can experience art. This initiative has been followed by other museums and academic institutions as well (Museo del Prado 2015). Likewise, 3D printers have been used for a multitude of aspects, such as for creating replicas to preserve the heritage of digital art works, such as for MOMO's Project, or to make musical instruments or short films.

In short, CIs are the ideal platform for new technologies to reach all audiences, due to their capacity for dissemination. It would be impossible to discuss every case of these recent innovations, since with each passing day comes a new invention in this field.

2.2 Successful Cases of the Use of Innovative Tools and Technology in the CI in the RPAS Field

If one technology could be said to have revolutionised the creative industries, in practically all areas, it would have to be drones. Drones are highly versatile pieces of technology, producing aerial images impossible to achieve by other means. They are undoubtedly the fashionable gadget of the moment and as such are evolving quickly due to their wide-reaching possibilities. Drones are in themselves a centre of emerging technological innovations, given the number of supported formats, which are also evolving, using both images as high-resolution frames or as video in multiple formats.

Job opportunities have started to emerge in the field of RPAS and there are now multiple companies that are producing work using aerial imagery, which has in turn made the legal regulation of these activities necessary. Legislation regarding drones is currently still being developed. According to Santamarina-Campos (2018), "[. . .] the European Agency of Safety Aviation (EASA) is continuing to work to provide a common regulatory framework to support the European competitiveness and leadership in the drone sector to deliver new employment and business opportunities".

Both the **photography** industry in aerial photography and **filming** present an indispensable resource for the creative industries (CIs). Due to the different heights that drones can reach and their ease of use, they are at present one of the sources of fundamental aerial images. The RPAS have revolutionised the taking of images throughout various sectors. Professional photography is highly competitive, and offering a new and different high-quality service could be the key to an individual's success.

The possibilities are endless, given that what was previously filmed with hand-held cameras, which obviously cannot fly, can now be filmed using drones (Image 1), and they are fast becoming a fundamental resource for the creative industry in all its sectors, at all levels, including professional. Because of their capacity to capture entirely different aspects from what we consider traditional images, there are already many photographic contests that reward the best images

Image 1 Drones used in
filming. Source: Clearhead

obtained from RPAS. An example of this is Skypixel Aerial Photography Prize, in
which images that are difficult to achieve by other means are obtained.

One of these fundamental sectors is **Architecture and Heritage**. This is one of
the sectors where drones have been most utilised and have already saved a lot of time
in carrying out tasks such as monitoring and inspecting the interiors of buildings, a
common task for architects. Within the field of heritage studies, drones are used for
tasks related to the conservation or maintenance of buildings or large structures, for
inspections of structural stability, etc., without the need to use scaffolds or other
safety equipment for images at a certain height. They are also used to help with tasks
in new construction sites, from inspecting the land, to mapping it, to taking measurements and coordinates. They save time and money for this type of operation,
where previously it was necessary to use expensive supplementary means or equipment, which often-required complicated assembly.

UAV Photogrammetry for Mapping and 3D Modelling is another of the disciplines that can be developed within the sector. When creating surveys of sites and
buildings, generating 3D spaces, with exact measurements is practically impossible
to achieve with other technologies.

In the **Arts and crafts** sector there is also a union between drones and graffiti. An
example of this is the creation of graffiti with RPAS, in the Painty by Drone® project,
developed by the design and innovation firm Carlo Ratti Associati. They completed
a project that, due to its great height, would have cost more time and more money
without the use of the drone. The project utilised four drones that carried sprays with
primary colours, in the CMYK system—magenta, cyan, yellow and black—as prints
have traditionally been made. The drones had built-in sensors to spray the walls with
colour and a central control system which allowed the drones to fly and to keep track

Image 2 Technology used
in advertising. Source:
Clearhead

of their position. These unmanned aircraft move within a network to prevent accidents. The drones painted drawings that were transmitted via an application. The work appeared on buildings and was a combination of pieces created by a single individual and of more people collected through crowdsourcing platforms" (Ratti 2017).

Advertising agencies provide a creative point of view to the marketing and branding strategies of their clients in order to promote and sell their products or services. These agencies are, above all, seeking to have more freedom of movement to explore new perspectives for their creative and artistic processes (Image 2). In an era where audiences are barraged with more advertising content than ever, marketers have to work hard in order to stand out. RPAS are literally unlocking an entirely new vantage point on the world, making aerial footage convenient, fast and cheap. They are used as innovative video tools and offer new perspectives.

An example of this would be the recording with drones for the **real estate** industry. Drones offer the possibility of registering a building from a 'polyangular' point of view, thereby offering a complete view of the property in a way that facilitates the task of advertising in an attractive and very visual way, of an exceptionally high quality, with access to hard to reach areas, and by minimising risks, costs and time.

Also, the use of promotional videos of sites, cities or localities and, fundamentally, intangible heritage, which are collected in video format, support uses and customs that form the identity of a certain group and that, not being material, serve as a conservative and promotional document to attract tourism. The drones have changed the way of filming these traditions and showing them to us from multiple points of view, adding plasticity and polyangularity to advertising videos, making them more attractive. An example of this can be found the ephemeral traditions, such as the Fallas of Valencia, Spain, recently named "Intangible Heritage of Humanity", whose transitory nature makes the filming and photographs taken from the drones become the heritage document itself that will preserve the experience historically. Some companies have recorded promotional videos with RPAS of the Fallas of the City Council of Valencia, which gives the possibility of visiting the now-disappeared monument, with a recording that serves as a documentary record for the exhaustiveness of its images (Airworksmedia 2014).

The use of drones applied to **fashion** has recently succeeded in the media, at the Dolce & Gabbana fashion show in Milan Fashion Week 2018. On this media catwalk, eight drones flew, each with a luxury bag from the Italian firm for the Fall Winter 2028/19 collection. A 3-minute choreography was orchestrated (CNBC 2018) in the middle of a music and light show, in the patrimonial area of the Oratory baroque di Santa Città Church in Palermo, Sicily.

Drones are fashionable and serve to advertise fashion. Drones are used to film or take pictures of this type of event, but in this case the function of the drones went further, forming part of the show itself. Dolce & Gabbana have used them to move mannequins, replacing human models, to present their new collection of handbags, combining fashion, with a luxury brand, new technologies and entertainment, to increase the media impact, generating an advertising strategy and pioneering the use of drones in this creative industry. The impact of the use of drones as mannequins has not only created a viral response from the mass media sector, but also from social media, becoming the protagonist of Milan Fashion Week 2018. They are a symbol of innovation and an introduction to the future (Image 3).

Movie industry This sector has greatly benefited from the use of drones, and has been incorporating them into their tools since the technology was first developed. In Hollywood (Flynn 2016), they are highlighted in the filming of films like The Wolf of Wall Street (2013), The Expendables 3 (2014), Chappie (2015), Specter (2015), Jurassic World (2015) or Captain America: Civil War (2016). The first one is cited by the author in 2012.

There are now productions that are completely shot using drones and film festivals exclusively for drones, such as Flying Robot International Festival (FRIFF), which is a celebration of aerial cinematography and since 2015 awards prizes in different categories such as:

- X-Factor & best in show
- News/documentary
- Narrative
- Landscape
- Extreme sports
- Freestyle/fpv
- Architecture

Image 3 Drones used in movies. Source: Clearhead

- Featuring drones
- Dronie
- Shaw reel (FRIFF 2017)

Television is a media for mass audiences and TV producers are known for their craft, originality and creativity. Television is present in most homes, and drones have managed to transform television production. Since re-launching programmes such as Planet Earth II with this new tool, which allows recording in 4K, it has been possible to achieve complicated panoramas that are impossible through other means. "No other filming method can start a sequence inside a building and end up at 400 f. altitude in one shot", says Ben Sheppard, managing director of Spider Aerial Filming. According to The Independent Spider, they have worked on programmes as diverse as Downton Abbey, 24 h in A & E and Channel 4's Dispatches to give the viewer a more complete immersion into the subject from the air (Newall 2016). The concept of taking television images has changed completely thanks to the perspectives that the drones achieve and the possibilities that they offer.

Performing arts includes theatre, dance, music (festivals), opera, magic, illusion and circus. These disciplines are often synonymous with movement, spontaneity and creativity.

There have been performances in which several drones have been choreographed to fly at the same time, by way of choreography, creating a "firework display", filling the sky with luminous RPAS, such as that in China in February 2017, organised by the company Intel®, presenting its prototype of drone Shooting Star®, a drone specifically designed to emulate fireworks with 4 billion colour combinations with LED lights, which combine colours and are built to be lightweight, in order to enhance safety in these type of spectacles, potentially making luminous pyrotechnics nearly obsolete (Morris 2017).

Another example, is the one used in the Mother of the Nation Festival, in March 2018, that showed a spectacular 500-drone light show in celebration of the "Year of Zayed" in Abu Dhabi, at the 100th birthday of the founding father of the UAE, Sheikh Zayed bin Sultan Al Nayhan. These 500 Intel® Shooting Star™ Drones create a new form of night time entertainment and storytelling—allowing creativity to come to life with the sky as our canvas, and flying lights as ink ("Signature Events—Mother of the Nation Festival" n.d.).

The drones "dancing in formation" is another of the topics proposed and improved in performing art. It is a mixture of human, robotic and animal movements; a space for music, dance, lighting and drones in which the show is creative and immersive. Tamás Vicsek and his research team from the Department of Biological Physics at the Eötvös University in Budapest created drones that communicate with each other directly and solve tasks collectively in a self-organised manner, without human intervention in the "Dancing with drones" project. The project was established with the aim of understanding the universal rules of the collective movement of animals to create an "autonomous robotic herd". The use of drones and robotics in this case becomes a spectacle, interacting with humans, shows technological progress and creative industries (Dancing with Drones 2015).

In the field of **music**, we find flying drones that become orchestral musicians, creating music and performing coordinated and synchronised musical pieces, including as a robot orchestra. KMel Robotics, a robotics company, designed RPAS capable of interpreting melodies and coordinating with each other (Waxman 2014). There are also hybrid musical shows between humans and drones, an example of which is "LOOP 60 Hz: broadcasts of The Drone Orchestra", musician John Cale and the architect Liam Young, in which RPAS fly over in performance mode, in games of lights, while the musicians interpret the music to the rhythm that they generate, creating an immersive visual spectacle with choreography and instruments, and generating live music.

The Cirque du Soleil has incorporated drones in its theatrical shows on multiple occasions, which have the possibility of adding to the magical effect in the circus environment, which give movement to inanimate objects, with choreography that generated special luminous effects, until creating the illusion of a ballet coordinated with precise algorithms (Rhodes 2014).

Another example of success in this field has been "Freedom Franchise", whose world premiere was in December 2017 at Miami Art Week, of the Studio Drift, together with BMW. It is a performance that used 300 luminous drones to create the effect of a flock of birds; studying the patterns of starlings led to software being specifically designed to integrate into drones, creating an immersive effect and reproducing animal patterns (Rhijnsburger 2017).

Another example of a successful case was the use of drones in the first theatrical performance of "Aerial Ballet", which was performed in 2015 in Laguna Beach, Beach, USA, in cooperation with development team Spark Aerial in San Diego. It was developed with seven Phantom 3 drones with integrated LED lights, that were synchronised to create artistic patterns. There is also the case of "24 drones" of Daito

Manabe, which combined the dance of drones with human interpreters, interacting with each other, generating emotional spaces (DJI 2015).

The **video game** industry uses registration of indoor environments for the development or exploration of game scenarios. These records can be used as a first step to creating more realistic video games. At the same time this industry seeks affordable and time saving solutions.

An example of successful use of drones for video games has been the UAVisuals company, along with Milestone, for a MotoGP videogame. The mission of the drones has been to carry out photogrammetric techniques and map the spaces to obtain geometrical measurements of the spaces in the images themselves. According to UAVisuals "there are more than 2 billion video game players worldwide, The eSports industry in particular is thriving. Revenue rose by 51.7%, to 401.8 million euros in 2016, and is expected to approach 1.22 billion euros by 2020. A huge part of this rise is thanks to advancements in technology and innovative new ways of acquiring imagery, such as with the use of drones. These advancements have meant game makers are able to produce products that are more realistic than ever before" (UAVisuals 2018).

Another example of success is Airhogs Connect Mission Drone, the videogame that brings together augmented and virtual reality, from a tablet with real drones that perform the tours over a real space, piloted as part of the videogame. The real drone movements are mixed on the screen with augmented reality, uniting real space and virtual space. It is "the first real drone immersed in an augmented reality gaming world" (Airhogs 2018). It is a video game that allows you to "submerge yourself in augmented reality as you fly through power rings, rescue civilians and battle aliens in this massive mixed reality world" (Airhogs 2018).

Moreover, the Official Star Wars drones have had great success among fans of the saga. Three models have been released so far: the Star Wars 74-Z Speeder Bike, the Star Wars TIE Advanced X1 and the Star Wars T-65 X-Wing Starfighter. "The drones themselves are tiny, featherweight models that are controlled via an included 2.4 GHz remote control. This remote also pairs with your smartphone over Bluetooth, giving you control via the free Propel Star Wars Battle Drones app (iOS/Android), which runs both training simulations and live battle tracking" (Cohen 2017). The release of the Millennium Falcon drone is predicted to coincide with the premiere of the Han Solo spin-off movie (Trenholm 2017). This example is a union of multiple areas of the creative industries, combining video games with cinema, generating epic battles that create interactive and filmic spaces.

3 Conclusions

After the detailed study of some of the most outstanding success stories of the use of new technologies and innovation in the creative industries, we can conclude that innovation is inherently linked to CI in the first place, given that despite the fact that they apply to other fields, the CIs are the main experimentation sector and, at the same time, the main platform for reaching the general public. The CIs linked to technologies arouse a great social interest, with which there is a transfer of innovation to the social fabric almost immediately, which is enhanced by social networks and the media. They have the capacity for instant dissemination because they adapt to society in an immersive way.

The CI, in addition to having the most diffusion of technology, at the same time generate a series of advances at an economic level, with entrepreneurship capacity, because they are integrated into social uses and because they generate benefits for around 1% of the world population, being 3% of world GDP. It is obvious that the advance of CIs is the advancement of culture, which has a direct impact on the production and distribution systems.

With this chapter, we want to point out the innovation applied to CI, but mainly the impact of the drones in the different areas of production, not only the ones we are accustomed to in sectors such as photography, video or cinema, in which they generate a visual documentation of great value, but also offering the possibility of being part of their own performances, video games, etc. In short, drones have proven to be the future and the possibilities they offer to CIs are endless.

References

Airhogs (2018) Aumented reality mission drone. Available via CONNECT.AIRHOGS. http://www. connect.airhogs.com/#wtb. Accessed 19 Jan 2018

Airworksmedia (2014) Fallas 2014. Available via AIRWORKSMEDIA. http://www.airworksmedia. com/cablecam.html. Accessed 8 Mar 2018

CISCAC (2015) Cultural and creative industries fuel global economy and provide 29.5 million jobs worldwide. Available via WORLDCREATIVE.ORG http://www.worldcreative.org/wp-content/ uploads/2015/12/COM15-0908_CCI_Fuel_Global_Economy_FINAL_EN.pdf. Accessed 20 Jan 2018

CNBC (2018) Drones carrying handbags kicked off the Dolce & Gabbana fashion show. Available via CNBC. https://www.cnbc.com/2018/02/27/watch-drones-at-dolce-gabbana-fashion-show. html. Accessed 5 Mar 2018

Cohen (2017) Propel star wars battle drones review. Available via DIGITALTRENDS. https://www. digitaltrends.com/drone-reviews/propel-star-wars-battle-drones-review/. Accessed 1 Mar 2018

Dancing with Drones (2015) About. Available via DANCING WITH DRONES. http:// dancingwithdrones.com. Accessed 7 Mar 2018

DJI (2015) Drones as art. Available via DJI. https://www.dji.com/newsroom/news/drones-as-art. Accessed 7 Mar 2018

Flynn S (2016) Drones in movies: 7 hollywood movies filmed with drones. Available via SKYTANGO.COM. https://skytango.com/drones-in-movies-7-hollywood-movies-filmed-with-drones/. Accessed 3 Mar 2018

FRIFF (2017) Flying robot international film festival. Available via FRIFF.COM. http://friff.co. Accessed 6 Mar 2018

Google (2015) Jacquard by Google®. Available via GOOGLE. https://atap.google.com/jacquard/ about/. Accessed 28 Jan 2018

Higgs P, Cunningham S (2008) Creative industries mapping: where have we come from and where are we going? Creat Ind J 1:7–30

Morris I (2017) Intel has drones that will make fireworks obsolete. Available via FORBES. https://www.forbes.com/sites/ianmorris/2017/09/19/intel-has-drones-that-will-make-fireworks-obsolete/#7189e12551db. Accessed 7 Mar 2018

Museo del Prado (2015) Touching the Prado. Available via MUSEODELPRADO.ES. https://www.museodelprado.es/en/whats-on/exhibition/hoy-toca-el-prado/29c8c453-ac66-4102-88bd-e6e1d5036ffa. Accessed 1 Mar 2018

Newall S (2016) How drones are transforming TV production. Available via INDEPENDENT.CO.UK. https://www.independent.co.uk/arts-entertainment/tv/planet-earth-ii-sir-david-attenborough-how-drones-are-transforming-tv-production-a6898336.html. Accessed 5 Mar 2018

Pezzini A, Konstantinou T (2013) Opinion of the European Economic and Social Committee on the communication from the Commission to the European Parliament, the Council, the European Economic and Social Committee and the Committee of the Regions

Ratti C (2017) Il progetto paint by drone. Available via CARLORATTI. http://www.carloratti.com/wp-content/uploads/2017/10/2017_10_01_AdV_-_Strategie_di_Comunicazione_pag.14.pdf. Accessed 18 Feb 2018

Rhijnsburger L (2017) World Premiere of Studio Drift's drone-based performance "Freedom Franchise" at Miami Art Week 2017. Available via DUTCHCULTUREUSA. http://www.dutchcultureusa.com/blog/4038/world-premiere-of-studio-drifts-drone-based-performance-freedom-franchise-at-miami-art-week-2017. Accessed 3 Feb 2018

Rhodes M (2014) Watch Cirque du Soleil's first magical experiment with drones. Available via WIRED. https://www.wired.com/2014/10/cirque-drones/. Accessed 6 Feb 2018

Santamarina-Campos V (2018) European union policies and civil drones. In: De Miguel Molina M, Santamarina-Campos V (eds) Ethics and civil drones. European policies and proposal for the industry. Springer, Heidelberg, pp 35–41

Snowball J (2016) Why art and culture contribute more to an economy than growth and jobs. Available via THECONVERSATION.COM. https://theconversation.com/why-art-and-culture-contribute-more-to-an-economy-than-growth-and-jobs-52224. Accessed 20 Jan 2018

Trenholm R (2017) Millenium Falcon to take off with 'Han Solo' movie. Available via CNET. https://www.cnet.com/news/millennium-falcon-drone-to-take-off-with-the-han-solo-film/. Accessed 5 Mar 2018

UAVisuals (2018) Drone mapping for MotoGP18 Video Game. Available via UAVisuals. http://www.uavisuals.com/portfolio/drone-mapping-for-video-game/. Accessed 7 Mar 2018

UNESCO (2017) Global report on cultural policies: focus on UIS data and analysis. Available via UNESCO.ORG. http://uis.unesco.org/en/news/global-report-cultural-policies-focus-uis-data-and-analysis. Accessed 5 Feb 2018

Wang H (2012) Research on information technology driven creative industries business model. In: Paper presented at the 2nd international conference on green communications and networks 2012 (GNC 2012), Chongqing, China, 14–16 Dec 2012

Waxman OB (2014) Watch flying robots play musical instruments. Available via TIME. http://time.com/74234/watch-flying-robots-play-musical-instruments/. Accessed 17 Jan 2018

Storyboarding as a Means of Requirements Elicitation and User Interface Design: An Application to the Drones' Industry

Ramón Mollá, Virginia Santamarina-Campos, Francisco Abad, and Giovanni Tipantuña

Abstract The goal of this chapter is to present the best practices and usage of storyboarding during the initial definition of multidisciplinary projects where partners from different backgrounds (engineering, arts, creative industries, etc.) collaborate to define the main user tasks to be implemented during the project. One of the challenges in this phase of the project is to be able to effectively communicate the ideas between the partners. Every background has his own concepts, technical language and procedures, and sometimes it is hard to convey in words the real meaning of an idea. It is even possible that different disciplines use the same tools, but have different names and different purposes. Storyboards are universally understandable and provide a common ground for sharing ideas and for discussing and discovering new points of view.

1 Introduction

Creative Industries (CIs) have used storyboarding for many years. It is a very useful and powerful tool for describing the content of a linear production, such as an animation film (Finch 2011). Storyboarding is a good collaborative technique, where all the members of the group can internalise the whole project and see the whole picture as well as small details. It also enables anyone to contribute his or her ideas effectively. Homogeneous teams of animators or audiovisual professionals regularly use this tool. Also, multidisciplinary multimedia production teams can work concurrently (Taylor 2013).

R. Mollá (✉) · F. Abad · G. Tipantuña
Instituto Universitario de Automática e Informática Industrial (ai2), Universitat Politècnica de València, Valencia, Spain
e-mail: rmolla@dsic.upv.es

V. Santamarina-Campos
Conserv. & Restoration of Cult. Heritage Department, Universitat Politècnica de València, Valencia, Spain

© The Author(s) 2018   83
V. Santamarina-Campos, M. Segarra-Oña (eds.), *Drones and the Creative Industry*,
https://doi.org/10.1007/978-3-319-95261-1_6

It is very common nowadays in the software development industry to design interactive systems based on user-centred design (UCD) processes. Any mid-to-large-sized modern software project typically involves multidisciplinary teams composed of technical and not technical stakeholders: end-users, clients, producers, designers and engineers. The different and complementary perspectives of the team members define the requirements of the system to be constructed. The ISO standard 24765:2010(E) defines requirement as "a condition or capability that must be met or possessed by a system, system component, product, or service to satisfy an agreement, standard, specification, or other formally imposed documents" (ISO/IEEE 24765-2010(E)). The software requirements specification document is the result of the requirement analysis stage of the project, and must consider all aspects of the application, ranging from user needs to non-functional requirements, such as safety, security, performance, reliability or latency requirements.

The methodology to obtain clear user requirements is key for the success of the project. This methodology has to:

1. Get all team members involved in the early stages of the system design.
2. Allow for the effective communication of the design ideas among all members.
3. Document the decisions in a way that all team members can understand and use them in an appropriate way in later stages of the process.

Understanding user requirements is an integral part of information systems design and it is critical to the success of interactive systems. However, specifying these requirements is not so simple to achieve. As specified in the ISO 13407 standard, UCD begins with a thorough understanding of the needs and requirements of the users. The benefits can include increased productivity, enhanced quality of work, reductions in support and training costs, and improved user satisfaction (Maguire and Bevan 2002). Requirements analysis is not a simple process. There are many methodologies for user requirement analysis that can be used to achieve these goals: stakeholder analysis, context of use analysis, video recording, focus groups, interviewing, scenarios and use cases, storyboards, etc. (Maguire and Bevan 2002).

Storyboards are a key and efficient means to communicate results of user needs analysis to the team members of a multidisciplinary group of professionals involved in user-centred software engineering (UCSE) projects (Haesen et al. 2009). Story-boards contain sketched information of users, activities, devices and the context of a future application.

An additional challenge when collaborating within a multidisciplinary UCD team is communication within the team without information loss. One missing link in most user-centred processes is an approach and accompanying tool to progress from informal design artefacts (e.g. scenarios) towards more structured and formal design strategies (e.g. task models, abstract user interface designs) without losing any information. Existing tools and techniques often require specific knowledge about specialised notations or models, and thus exclude team members not familiar with these notations or models. Furthermore, functional information may be missing in informal design products, while structured design results may not always contain all non-functional information. Storyboards are a comprehensible notation that allows these shortcomings to be overcome (Haesen et al. 2016). Storyboarding has also proved to be a good methodology for developing interactive multimedia applications such as video games (Lambert and Jacobsen 2015).

A storyboard graphically represents a sequence of actions or events that the user and the system being designed go through to achieve a task. A scenario is one textual story about how a product may be used to achieve the task. It is therefore possible to generate a storyboard from a scenario by breaking the scenario into a series of steps which focus on interaction and creating one scene in the storyboard for each step (Rogers et al. 2011). The purpose for doing this is twofold:

1. To produce a storyboard that can be used to get feedback from users and colleagues.
2. To prompt the design team to consider the scenario and the product's use in more detail.

After this process is finished, the technical team can process the stories and develop a draft that is focused on the way the user will interact with a hypothetical application. This application implements the interactions portrayed in the storyboards. After the technical team validates this draft, the rest of stakeholders can evaluate it. Stakeholders can provide feedback with new suggestions, objections or confirmations. The technical team process this feedback and produce a prototype of the application that will have all the main features of the definitive application, taking into account the suggestions of the stakeholders. A number of cycles of design-test-implement are performed in order to refine the visual aspect of the application, workflows, ergonomics, etc. before developing the final system.

2 Using Design Thinking in AiRT

The main goal of the AiRT project is to provide the European creative industries with a new tool that will enable them to offer new services and to grow in the international market. To achieve this objective, we have designed the first RPAS that is specially designed for professional indoor use (Santamarina-Campos et al. 2018).

Inclusive and participatory methodologies and work tools have been used in order to achieve the specific results and objectives of the project. These methodologies have allowed collaboration and communication both internally (between the consortium) and externally (with the end-users) of the project. For this reason, Design Thinking methodology (Both 2009) has been used to ensure that:

1. Results are aligned with the values of the creative industries (García and Dacko 2015) from the initial stages of innovation processes.
2. Communication between the interdisciplinary team that makes up the consortium (composed of engineers, economists, artists, creatives, etc.) is effective.

This methodology (see Fig. 1) is centred on the user experience, focusing on the design process rather than the final product. It allows the convergence of different fields combined through "radical collaboration", with the common goal of

Fig. 1 Methodologies used in the design of the Ground Control System (GCS) software. Source: Own elaboration, adapted from Both (2009)

implementing tools that enable new flows of thought based on intuition, critical thinking and creativity (Brown 2009).

Thereby, storyboarding can be found in the intersection between narrative, visual thinking, and design thinking (Beckman and Barry 2007; Wikström and Berglund 2011).

3 Methodology Development at AiRT

Design Thinking (Both 2009) has allowed the consortium to be connected to the experience of the creative industries. Creative Industries professionals have participated in the identification of needs and have interacted with the prototype during demonstrations. The aim was to learn from user feedback. Design Thinking is a suitable methodology for this interdisciplinary consortium, formed by engineers, managers and creatives, and from the integration of experts from 13 sectors of the European creative industry (Santamarina-Campos et al. 2018). The five steps are analysed and developed below.

1st Phase. Empathise

The project began with the analysis of the needs of the creative industries. This analysis was carried out with end-users, which allowed us to obtain consistent results. The technique of the Stakeholder Map was used to identify the potential users of the product. It allowed us to have a clear image of the users who had to intervene and participate in the analysis. Based on the stakeholder map defined by the consortium, the key informants who participated in the definition of the features of the system were identified through three focus groups in Spain, Belgium and the United Kingdom. The dynamics were prepared by means of a previous Documentary Investigation around the requirements related to aerial filming and photography, the use of drones and the security and data protection problems derived from them. The key informants chosen at this stage came from 13 different sectors of the creative industries and they participated in the entire project process.

2nd Phase. Synthesise

An analysis of the needs of the CIs and ethical and risk issues was carried out based on the information extracted in the previous phase. Qualitative Content Analysis, Social Network Analysis (SNA), and manual coding and categorisation of qualitative data were the methods used for processing the gathered information. The first step was to identify the real needs of the end-users. These needs drove us to define possible key solutions that provided added value and let us obtain an innovative result. The results obtained at this stage were the basis of the idea of the AiRT system.

3rd Phase. Ideation

The target of the ideation stage is that both CIs and creative professionals, who are part of the consortium, can define the functionalities to be implemented in the AiRT system. The AiRT system is composed of an RPAS that is driven automatically from land by a Ground Control System (GCS) software. The GCS runs on a standard tablet. Written scripts (Fig. 2) were elaborated, starting from the identification of the needs carried out in the initial phase, together with the specifications included in Annex 1 (part A) of the DoA,[1] in the Grant Agreement no. 732433.

These written scripts were subsequently transferred to graphic scripts (Storyboards) (Fig. 3) that represented the use of the AiRT system in different creative scenarios. Storyboarding was one of the main techniques used during the process of requirements elicitation. It allowed great visual and plastic content. Storyboarding eases creative and analytical thinking while facilitating communication between internal groups of the consortium. Innovative and feasible solutions arise around

---

[1]Description of the Action.

WP5 Task 5.1 Software adaption - user interface programming
D5.1: End-user friendly adapted software

STORYBOARD – AiRT SYSTEM

SYNOPSIS

A young filmmaker is filming his first feature film. One of the main scenes of his first work will be filmed inside an old textile factory of the early twentieth century. The chosen space as a stage, is a large abandoned industrial building of 1000 m² of rectangular porticoed plant. It presents two rows of fine columns, typical of textile factories of that time, that run along the length of the room dividing it into three equal sections. The deck is flat, with exposed beams, with a height of 4.5 meters. It features small sized spans located at the top of one of the sides of the ship, allowing controlled entry of natural light, creating an intimate atmosphere. It features small sized spans located at the top of one of the sides of the industrial building, which allows controlled entry of natural light, creating an intimate atmosphere. The scene to be shot represents the game of two five-year-old children between the columns. This scene will not only represent the film, but also be used for the design of the poster, thus it will be necessary to take photographs. It tries to obtain a free movement of the camera,

which accompanies the freedom of the children's movement. Therefore, a not invasive tool is needed, so that the children can move with freedom and spontaneity, that allows to obtain continuous and clean shots of the space. That is why the use of rails, cranes, and other auxiliary means that may obstruct the space and limit the free movement of the camera is discarded. On the other hand, the recording should be done in the blue hour, so the shooting time will be very brief and will not allow the repetition of shots.

Technology transfer of Remotely Piloted Aircraft Systems (RPAS) for the creative industry - AiRT
www.airt.eu

WP5 Task 5.1 Software adaption - user interface programming
D5.1: End-user friendly adapted software

VISUAL PLANIFICATION

Time Line	Type of shot	Scene/ Utility/ Actors	Shot description	Dialogues
1	Long shot zenith in the back of the characters	Interior textile factory of 1900. Natural lighting through the windows, 13:00 hours. Director Cameraman Pilot Assist Drone AiRT	In the abandoned textile factory, the director, the Cameraman and the Pilot Assist meet the day before the shooting of the scene. They are in the centre of the hall, around the drone AiRT. Each person has some physical trait that characterizes him. On their shirts, the role they play in the film is indicated.	**Text at the bottom of the bullet:** *Day before the filming of the main scene of the film, the director, the Cameraman and the Pilot Assist meet at the stage chosen to prepare the filming.* **Balloons:** Cameraman: *The location is perfect: a labyrinth of columns that will accompany the children's game.* Director: *The shot with the children can not be repeated, it has to be perfect at first try, to collect the spontaneity and freedom of movement of the children. Remember that we will film in the blue hour.* Cameraman: *So, it is fundamental that we have very clear the path of the camera.*
2	Long shot zenith head on the characters	Director Cameraman Pilot Assist Drone AiRT	In the long shot, slightly zenith of the industrial building, the characters are seen from the front and the anchors are highlighted. In a corner, a close-up of a tablet is shown with a finger marking "calibrate". The Pilot Assist presses the calibration button on the Tablet and the drone moves in the "Z" axis half a meter up, and makes a 360 ° turn without barely moving from the take-off place.	**Text at the bottom of the bullet:** *AiRT is ready, the application shows at the top of the window a timeline, which indicates sequentially the necessary steps to reach the final flight. At the moment the timeline runs to the first milestone indicating that the equipment is ready, waiting to comply with the next step that will be the calibration.* **Balloons:** Pilot Assist: *... So, first thing, let's prepare the equipment, I've already put 4 anchors for the iPS. It calibrates in a second. Look ... can you see it? This green flashing button indicates that everything is ok and we can do the calibration, press here and ... ready! And now we can start with the mapping.* Director: *Perfect! How long will it take you to do the mapping?* Pilot Assist: *It will be fast, because it does not have much height. In three passes it will be ready, one pass for a meter and a half.* Cameraman: *Does the mapping have only one flight option?* Pilot Assist: *No, it has three, depending on the support you need. Look, take the Tablet and you'll see how I do it.*

Technology transfer of Remotely Piloted Aircraft Systems (RPAS) for the creative industry - AiRT
www.airt.eu

Fig. 2 Examples of written scripts that describe a hypothetical storyboard scenario. Source: Own elaboration

Fig. 3 Collection of the graphic script covering a storyboard to define possible innovative solutions after identifying real needs of the end-users. Source: own elaboration

storyboards. The use of this tool has allowed us to fulfil one of the main premises of the creative process of Design Thinking, "Show, don't tell" (Plattner 2010). This means that it is important to communicate the vision in an impacting and meaningful way by creating experiences, using illustrative visuals and telling good stories.

The use of this tool favoured expansive thinking. Programmers could communicate the main ideas that were drawn from the previous phase directly and more clearly, without value prejudices. The purpose of the use of storyboards was to help the creatives involved in the consortium to transfer real scenarios to the design process of the AiRT system. These scenarios created possible indoor spaces that would reflect the identified needs in different creative spaces.

An heuristic analysis of the main programmes for flight plans based on mesh or mosaic was performed in this phase. This analysis was also done for the best mapping and photogrammetry programmes available on the market. The goal of this study was to analyse existing solutions similar to our product in order to have a more complete perspective of issues around the usability and final design of our tool.

To conclude this phase, a storytelling of the history of the project was elaborated, with the target of informing the creative industries and other sectors of the potential of the tool.

4th Phase. Prototype

In this phase, the ideas expressed in the previous stage were materialised. Bear in mind that the development of prototypes is not simply a way to validate ideas, but it

Fig. 4 Participation, action
and research interoperation
generates PAR
methodology. Source: own
elaboration

is an integral part of the innovation process (Plattner 2010). Therefore, this phase
carries out many other implicit goals. The AiRT RPAS is a complex system that
involves a high-end Indoor Positioning System (IPS) technology that was in beta
stage of development. AiRT RPAS can be seen as a real-world test bench for the IPS.
A new drone was also developed in order to match the requirements of an indoor
drone with support for professional cameras. This drone had its own standard Flight
Control System (FCS). Finally, the Ground Control System (GCS) connects wire-
lessly to an On-board Control System (OCS). During this phase, all components of
the system, FCS, OCS, GCS and IPS were integrated. Meanwhile, a software
prototype that implemented the functionalities of the AiRT system was developed.
This software runs on the GCS. It has to communicate bidirectionally with the server
process running on the OCS. The design of this software was based on prioritised
requirements, with the goal of making the solutions visible. The Human-Computer
Interface of the software (user interface) was based on the heuristic analysis of the
information extracted from the storyboards done in the previous stage.

Potential improvements were identified, but they were relegated for later developments, after users had a chance to interact with the product and provide feedback.

5th Phase. Tests
During this stage, end-users in the three participating countries tested prototypes from
a selection of scenarios relevant to the creative industries. The objective of this stage
was to involve users in the identification of failures or in the contribution of new
improvements, through the **Participatory Action Research** tool (PAR) (Chevalier
and Buckles 2013). PAR focuses on the effort to integrate three main aspects:
Participation of all the stakeholders, Action as an engagement with experience and
history and Research to extract the knowledge (Fig. 4). The purpose of this type of
technique is to obtain relevant data from key informants that allow subsequent
interpretation and analysis of the facts from the experiences (Santamarina-Campos

et al. 2017). In parallel, the dynamics were recorded with the objective of using the "covert observation technique" (Bryman 2016), through the analysis of the filming of the sessions, and then interpretation was done using qualitative data analysis software.

A workshop is also another activity for making the tool known to the CIs among other sectors. The aim is to bring AiRT closer to the market. Finally, two storytelling lines will be developed:

1. One will show the whole process of the project, from the ideation phase to test.
2. Another one for commercial purposes only.

4 Requirements Elicitation and User Interface Design in AiRT

Requirements analysis is a set of activities performed to determine the needs or features that a new product should have. It is an important stage in any software engineering project. Requirements analysis is critical to the success of a software project. It helps to identify business needs or opportunities. It defines the functionality of a programme to a level of detail that is high enough to perform a system design. The requirements should be documented. Every requirement has to be:

1. *Feasible*: it can be accomplished by a computer program.
2. *Measurable*: it is computable and can be translated into numbers in some way.
3. *Validatable*: the development team has to test and check that what the programme does is what the requirements states.
4. *Traceable*: there is always a user need that justifies any action that the programme does. On the other side, the programme does not do anything that has not been requested by the customers.

Requirements analysis includes three types of activities that have been used for this project:

1. *Eliciting*. Requirements gathering or requirements discovery are performed by analysing documentation of the business process and by interviews with the stakeholders. After these meetings, the technical team generates an intermediate documentation consisting of some user stories and storyboards.
2. *Analysis*. A phase where the team determines whether the informal requirements obtained in the previous stages are clear, complete, consistent and unambiguous. Sometimes these requirements cannot meet everyone's expectations. So, some decisions have to be taken in order to resolve any apparent conflicts.
3. *Recording*. This phase is in charge of documenting the requirements including a summary, natural-language documents, storyboards and anything else that can affect any other area of the software production: data models, communication models, process specifications, etc. The software requirements specification

document is the distillation product that emanates from all the requirements analysis process.

The first activity, eliciting was performed by using focus groups as seen in the 1st phase, Empathise. The second activity, Analysis, was made by using storyboards, as seen in the third phase of Ideation. The third activity, Recording, tried to detect in a first step the main areas where the requirements could be assigned. We used collaborative technologies for the development team. This allowed us to perform asynchronous collaboration, effective comments exchange, the integration and unification of definitions and the reduction of redundancy.

Although the goal of this phase was to obtain a list of user interface (UI) requirements, during the analysis many other requirements not related directly to UI appeared. These involved programming and communications requirements that emanated from the analysis of the user interface and software functionality analysis. The latter types of requirements are especially important for this project since they involve all the three development teams: software, IPS and drone development. These requirements are for internal use only and they were reserved for later use by the development team.

The storyboard scenarios, as described above, are descriptions of practical situations where professionals use the drone for recording needs. Most user requirements are obtained from storyboards. Several storyboards were proposed and catalogued.

Requirements have to be traceable. This means that every requirement has to be connected to the storyboard where it emerged, allowing for the quick location of both requirements and storyboards. We provided a unique identifier for each requirement and for each storyboard. References to storyboards follow the syntax:

$$SourceType + \mathsf{XX} + \text{":"} + page$$

where

SourceType is the type of document where the storyboard is described. For instance, UC as an acronym for User Case, but there are other types of documents used during the analysis stage.
XX is the number of the storyboard.
page is the page number where the requirement appears inside of the storyboard.

After analysing the storyboards, the requirements are compiled on a list. This list may change during the development phase due to user feedback or the discovery of new emerging functionality that can complement or clarify some ambiguous requirements. After some time, and with the whole consortium's agreement, the requirements list is frozen. This list defines the initial functionality of the system. The goal of this step is avoiding requirements creep, that is, uncontrolled growth in the project's scope during the development phase.

As the requirements were identified, they were also classified. Related requirements were grouped into categories. Table 1 shows the categories used in the project.

Table 1 Definition of categories

Category name	Acronym	Description
Safety	SAFE	Security and safety. The drone has to fly and return home without incident or endangering its integrity or people in the flight area or nearby.
Power	POWE	Features regarding energy spent by the system and power up/shutdown procedures.
Looks and design	LOOK	Appearance of the drone. Hardware design: housing, careen, etc.
Pre-production	PRE	Anything happening before the CALI step.
Calibration	CALI	A phase previous to RECO where the IPS bases are set and recognised by the system in order to allow a precise localisation of the drone.
Recognition	RECO	Flight phase where the flight environment is scanned in order to provide an environment for helping during the design of the recording flight plan.
Flight plan	FPLA	Process of creating the flight plan in the virtual environment map: definition of waypoints, path edition, etc.
Recording	REC	Execution of the selected flight plan over the real space for recording the video or taking photos.
Post production	POST	Anything happening after the REC step.
In flight	FLY	Applies whenever the drone is flying (for any purpose).

Source: own elaboration

Each category represents a state of the application workflow, or an area of the system.

A requirement can be assigned to:

– A single category. For instance, a requirement such as "the initial calibration of the positioning system must be performed automatically" has been assigned to the category CALI.
– Several categories. For instance, a requirement such as "the drone must not collide with any wall when performing the environment 3D mapping" may be assigned both to categories RECO and SAFE. Whenever a requirement is classified in more than one category, one of them is chosen as the primary.

The description of each requirement is organised in several fields (see Fig. 5):

– *Identifier* (ID). It provides an identity (name) to the requirement. It follows the convention of RXXX where R is the short for "requirement" and XXX is the number assigned to this current requirement. Requirement IDs are never reused. Therefore, when two different requirements are found to overlap, one of the two IDs is deleted and never used again, and its requirement is assigned to the remaining IDs. Therefore, the final list of requirements will have gaps in the requirements IDs. This process helps with the requirement traceability.

ID	R004
Category	FPLA
Description	The application has to know or ask for the predefined operational limits of the drone (maximum speed, maximum acceleration, etc.), the gimbal and the RCAM The application has a file or a configuration which can be used to load these values. The same file or configuration should be loaded into the server on the OCS.
Actors	Client application, OCS
Source Storyboards	UC03:1, UC04:1, UC06:1, UC08:1, UC08:2, UC09:2, D33:7, D33:17
Precondition	This process starts when the application starts.
Comments	All of the drone main telemetry parameters are sent from the FCS to the OCS. These data are sent back to the GCS in order to be shown at the UI if required.
Responsible	AEROTOOLS
Version	1.0

ID	R026
Category	SAFE
Description	The user can specify the minimum safety distance to obstacles. These values should be taken into account to perform the flight at "Mapping" step or at "Recording" step.
Actors	Client application, OCS, FCS
Source Storyboards	UC00:9
Precondition	Define settings values is the first step before the drone takes off.
Comments	The values introduced by the user have to be in a valid interval. The safety distance to obstacles (walls) is needed to calculate the different heights at which the mapping could be done. These requirements have to be sent to the OCS in order to warn FCS to change trajectory in case security is compromised.
Responsible	UPV
Version	1.0

Fig. 5 Examples of requirement definition. Source: own elaboration

- *Category*. The category assigned to the requirement. It follows the code provided above.
- *Description*. An explanation of the goal of the requirement. This list contains high-level requirements that, during development, may be subdivided into requirements that are more detailed.
- *Storyboards*. This field enables us to relate each requirement with the storyboard it derived from, and therefore to trace it. There must be at least one customer need that justifies any function implemented in the system. There are many internal technical requirements that were not explicitly mentioned by the customers

because of the customer's lack of technical knowledge. These requirements have to also be traceable and have to appear in the list, although they are secondary requirements made by the development team. In any case, there is always a reference to the code assigned to the storyboard that justifies the current requirement.

– *Responsible*. It is the team or the person in charge of developing this requirement. They have to supervise or execute the correct implementation of the requirement.
– *Comments*. They allow expanding or refining a bit more the description of the requirement. This field is not mandatory.
– *Version*. It indicates the assigned version of the software when the requirement will be implemented. Current version of the software is V1.0. If the requirement is assigned to the first version of the drone, the requirement is mandatory. It has to be developed within the project scope. If the requirement is not important or it is not clear if industry will demand it urgently, it is set to a later version. If there is time enough to complete it, perhaps it will be developed and integrated in the current version. If the requirement demands a lot of work and it is not clear that the development could complete the requirement in time, it will be assigned to V2.0 for the future development of the drone.

5 Conclusions

Requirements Analysis is a complex process that involves users and developers during most of the software development cycle. At the beginning of the project, requirements are obtained using different techniques with all the stakeholders. Later, during development, requirements are refined and adapted to the development evolution, and it is necessary to keep users updated and take their feedback into account.

We obtained and analysed the real needs of our users through three focus groups. We found that current indoor drones do not provide the features that creative industries demand, and thus we were able to detect many possibilities for the AiRT system. At this point, the use of storyboards made it possible to simplify and add a narrative and an appropriate context that eased the design process.

Storyboards provided visuals for the consortium's developers with real uses and potential user experiences, based on the identification of needs. Storyboards were an invaluable tool for developing the requirement list that describes the specific functionalities to be implemented in the AiRT system, and specifically in the GCS software. The graphic scripts helped to promote communication between software developers and the consortium's creatives.

Many GUI requirements were specific for this system since there is no comparable product on the market. CI users expressed their requirements from previous experiences with drones in outdoor scenarios. They also expressed:

– their problems in moving auxiliary equipment for filming indoors, especially in inhabited spaces or places with heritage value (Informant 6 and 8);
– invasive aspects involved in the use of auxiliary means such as lifting platforms, travel or sliders (Informant 1);
– the time-dependent condition of filming or photographing indoors with natural light and the difficulty of repeating identical shots (Informant 19);
– the limitations of auxiliary means of obtaining special footage (Informant 20); and
– experiences of using indoor drones, highlighting the need to incorporate an indoor positioning system and safety measures (Informant 13).

On the other hand, although current technology for outdoor drones is mature, this technology does not match the more demanding indoor requirements. Therefore, we had to push this technology to its end in order to reach the minimum specifications considered essential for the CI.

After the analysis of the user requirements, there were some additional technical requirements that arose because of the demands of future users. These requirements were mainly about:

– how the device is introduced in the industry's workflow;
– security specifications, some mandatory, some complementary;
– error control when flying; and
– other difficulties that appeared when implementing in restricted hardware platforms, for instance, the amount of geometry information that can be stored by the OCS depending on the memory available, transfer speed depending on the Wi-Fi network, rendering limits of the tablets used for the project, etc.

Putting the specifications together, we found that some were not feasible in the current state of drone technology. Others were also not feasible because they go beyond the project objectives and thus cannot be implemented with the given timeline and resources. So, some requirements have been postponed for later versions.

These specifications may still change, especially as beta versions of the system start to be used by the final users. Then they will realise the potential of the indoor drones and start to suggest more changes concerning the GUI.

References

Both T (2009) Bootcamp Bootleg. https://dschool.stanford.edu/resources/the-bootcamp-bootleg
Brown T (2009) Change by design: how design thinking can transform organizations and inspire innovation. HarperCollins, New York
Beckman SL, Barry M (2007) Innovation as a learning process: embedding design thinking. Calif Manage Rev 50(1):25–56
Bryman A (2016) Social research methods. Oxford University Press, Oxford
Chevalier JM, Buckles DJ (2013) Participatory action research: theory and methods for engaged inquiry. Routledge, Abingdon

Finch C (2011) The art of Walt Disney: from Mickey Mouse to the Magic Kingdoms and beyond. Harry N. Abrams, New York

García R, Dacko S (2015) Design thinking for sustainability. In: Swan S, Griffin A, Luchs MG (eds) Design thinking: new product development essentials from the PDMA. Wiley, New Jersey, pp 381–400

Haesen M, Vanacken D, Luyten K, Coninx K (2009) Supporting multidisciplinary teams and early design stages using storyboards. In: Jacko JA (ed) Human-computer interaction. New Trends, HCI 2009. Springer, Berlin, pp 616–623

Haesen M, Vanacken D, Luyten K, Coninx K (2016) Storyboards as a Lingua Franca in multidisciplinary design teams. In: Collaboration in creative design. Springer, Cham, pp 211–231

ISO/IEEE 24765-2010(E) (2010) Systems and software engineering, vocabulary. https://www.iso. org/standard/50518.html. Accessed 16 Mar 2018

Lambert D, Jacobsen M (2015) Building digital video games at school: a design-based study. In: Preciado BP, Takeuchi M, Lock J (eds) Proceedings of the IDEAS: designing responsive pedagogy conference. Werklund School of Education, University of Calgary, pp 32–42

Maguire M, Bevan N (2002) User requirements analysis. A review of supporting methods. In: Baeza-Yates RA, Montanari U, Santoro N (eds) Proceedings of IFIP 17th world computer congress. Kluwer Academic, Montreal, pp 133–148

Plattner H (2010) Stanford D. School Bootcamp Bootleg. Available at Institute of Design Thinking at Stanford. http://dschool.stanford.edu/wp-content/uploads/2011/03/BootcampBootleg2010v2SLIM. pdf. Accessed 15 Feb 2018

Rogers Y, Preece J, Sharp H (2011) Interaction design: beyond human-computer interaction. Wiley, New York

Santamarina-Campos V, De-Miguel-Molina B, Segarra-Oña M, De-Miguel-Molina M (2018) Importance of indoor aerial filming for creative industries (CIs): looking towards the future. In: Katsoni V, Velander K (eds) Innovative approaches to tourism and leisure. Springer, Cham, pp 51–66

Santamarina-Campos V, Martínez-Carazo EM, Carabal-Montagud MA, De-Miguel-Molina M (2017) Participatory action research (PAR) in contemporary community art. In: Santamarina-Campos V, Carabal-Montagud MA, De-Miguel-Molina M, De-Miguel-Molina B (eds) Conservation, tourism, and identity of contemporary community art. Apple Academic Press Taylor & Francis, New Jersey, pp 105–136

Taylor A (2013) Design essentials for the motion media artist: a practical guide to principles and techniques. Taylor & Francis, New York

Wikström A, Berglund K (2011) A narrative approach towards understanding innovation. In: Proceedings of 18th international product development management conference, 2011

Usability and Experience of the Creative Industries Through Heuristic Evaluation of Flight Software for Mapping and Photogrammetry with Drones

Virginia Santamarina-Campos, María-Ángeles Carabal-Montagud, María de-Miguel-Molina, and Blanca de-Miguel-Molina

Abstract This work presents a heuristic analysis and evaluation of the main programs of mesh or mosaic flight plans for mapping and photogrammetry. The objective of this study was to identify the best designs linked to certain factors and usability elements to avoid errors and identify opportunities for optimization in the design of the Ground Control System (GCS) software. The GCS, through a graphical user interface (GUI), provides an advanced indoor navigation system for the drone, which was developed within the framework of the H2020 European Project AiRT (Arts Indoor RPAS Technology Transfer) (Definition of AIRT, *chiefly Scottish*: compass point).

1 Introduction

SMEs represent 85% of the stakeholders in the creative industry sector in Europe. They face competition from large companies and often encounter the challenge and need to adapt cutting-edge information and communication technologies (ICTs) with limited resources. For this reason, ICT tools, and in particular technological innovation, are essential for increasing the competitiveness of creative industries, because they expand the creative possibilities and improve the efficiency in all sectors (European Commission 2015).

V. Santamarina-Campos (✉) · M.-A. Carabal-Montagud
Conserv. & Restoration of Cult. Heritage Department, Universitat Politècnica de València, Valencia, Spain
e-mail: virsanca@upv.es

M. de-Miguel-Molina · B. de-Miguel-Molina
Management Department, Universitat Politècnica de València, Valencia, Spain

© The Author(s) 2018
V. Santamarina-Campos, M. Segarra-Oña (eds.), *Drones and the Creative Industry*,
https://doi.org/10.1007/978-3-319-95261-1_7

This work is part of the European Project AiRT,[1] "Technology transfer of remotely piloted aircraft systems (RPAS) for the creative industry", the overall objective of which is to provide European cultural and creative industries with a new tool, enabling them to offer new services and grow in the international market. To achieve this goal, we have designed an RPAS especially for indoor professional use (Santamarina et al. 2018). The main innovations associated with this tool have focused on the integration of the following:

1. An indoor positioning system (IPS), based on ultra-wide-band (UWB) technology. This system allows safe navigation of drones indoors, providing movement and positioning information to a few centimetres on all axes. This can be achieved due to an improved update rate of up to 240 Hz, a four-antenna approach, and adapted positioning algorithms.
2. Software that allows the reconstruction of the indoor space (3D model) where the recording will take place. This allows the creative space to be analysed beforehand and the flight plan to be designed and executed in a safe way.
3. GCS software, which provides an advanced navigation system through the GUI. It also includes active safety measures, with the possibility of planning the flight and designing the audiovisual project.
4. A drone with ready-to-use commercial (COST) element integration on an aerial platform that fulfils the latest passive security measures at a reduced price, integrating a professional camera that operates with 360° rotation both at the top and at the bottom of the drone.

In addition, the value proposition covers a wide range of issues relevant to the target segment of the creative industries: price, security, operation, usability, and onboard technologies. All these aim to offer additional services and are designed to provide a positive customer experience.

2 Methodology Developed in the Design of the Ground Control System Software

The aim of the AiRT project is to generate an innovative tool that focuses on effectively understanding and solving the real needs of the creative industry. In the specific case of the design of the GCS software and GUI, ISO 13407:1999 provides a guide to achieving quality in use by integrating iterative tasks into user-centred design (UCD). UCD considers this as multidisciplinary work, which includes human factors and knowledge as well as ergonomic techniques, with the objective of optimizing the effectiveness and efficiency of the work environment and neutralizing the possible adverse effects of its management (ISO 1999).

---

[1]The project has received funding from the European Union's Horizon 2020 research and innovation programme under grant agreement no. 732433. Reference: H2020-ICT-2016-2017.

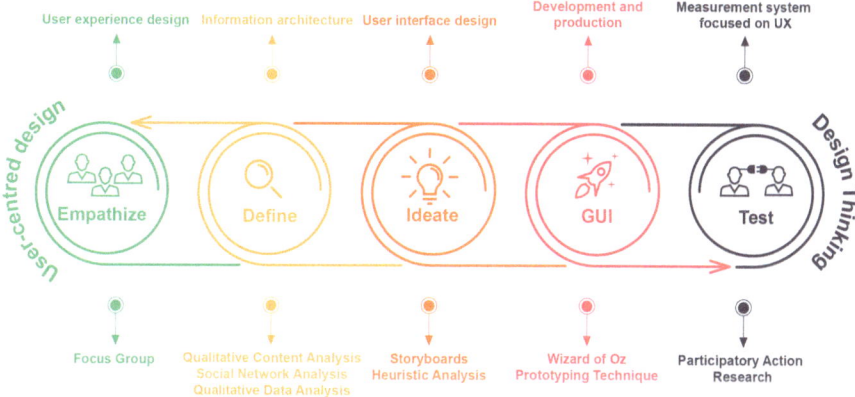

Fig. 1 Methodologies used in the design of the GCS software and GUI. Source: Own elaboration, adapted from Both (2010)

To achieve significant innovations, it is necessary to know the end-users and care about their lives (Both 2010). To complement the UCD, the design thinking methodology (Fig. 1) was implemented with the aim of generating a feedback process with the cultural and creative industries through collaborative, participatory, and creative work. This method is mainly composed of five non-linear phases, called empathize, define, ideate, prototype, and test. These lead to a solution that meets the objectives of the end-users and is technologically feasible and commercially viable (Both 2009).

2.1 First Phase—Empathize—and Second Phase—Define

In the specific case of the development of the GCS software and the GUI, in the first phase of the project, known as empathize, three focus groups were carried out in Spain, the United Kingdom, and Belgium, from which information was obtained on thirteen different sectors of the creative industries. This technique constituted an effective qualitative tool to discover the desires, motivations, values, and experiences of our users (Hinton 2004). As a result, during the define phase, a need analysis was carried out through qualitative content analysis and social network analysis (SNA). Therefore, manual coding and categorization of qualitative data (Santamarina et al. 2018) were applied, which formed the basis for the analysis of the functionality of the program executed in the GCS, the functionality of which is provided to the human end-user through the GUI.

2.2 Third Phase—Ideate

In the third phase, based on the synthesis of the information obtained in the focus groups, written scripts were prepared and then transferred to storyboards representing the possible functionalities of the GCS software in different creative scenarios. These helped to communicate the main ideas and needs more clearly. Storyboards, also termed "presentation scenarios", are image sequences that show the relationship between the user's actions or inputs and the results of the system, making it easier for the design team to understand them (Maguire and Bevan 2002). This technique can help to control the process of creating solutions and to identify different variables to break down large problems into smaller ones that can be evaluated and solved better (Both 2010). In this way, from the storyboards, the requirements were extracted, which allowed us to obtain, on the one hand, the concrete functionalities to be implemented in the GCS software and, on the other hand, the definition of elements related to the usability aspects of the GUI.

At the same time, the requirements of the GCS software were defined based on the needs of the creative industries. A documentary investigation of 29 flight plan software programs was carried out with the objective of identifying the solutions available in the market that are similar to our product to gain a more objective perspective on the usability aspects and final design of the GUI. The main tool used for the development of this analysis was a heuristic evaluation (Molich and Nielsen 1990), allowing the analysis of the main usability components of flight plan software by experts and focusing the study on the variables ease of learning, efficiency, quality of being remembered, effectiveness, and satisfaction (Nielsen 2012).

2.3 Fourth Phase—Prototype

Once the heuristic evaluation of the flight plan software available in the market had been completed and the design elements extracted from the storyboards had been defined (the user interface of the client application), an iterative design process was

Fig. 2 Paper prototype of the clients' UI. Source: AiRT project

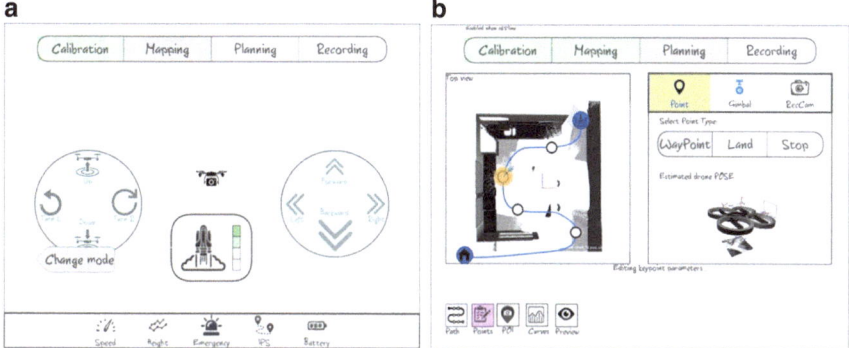

Fig. 3 (**a** and **b**) Graphical user interface layout design. Source: AiRT project

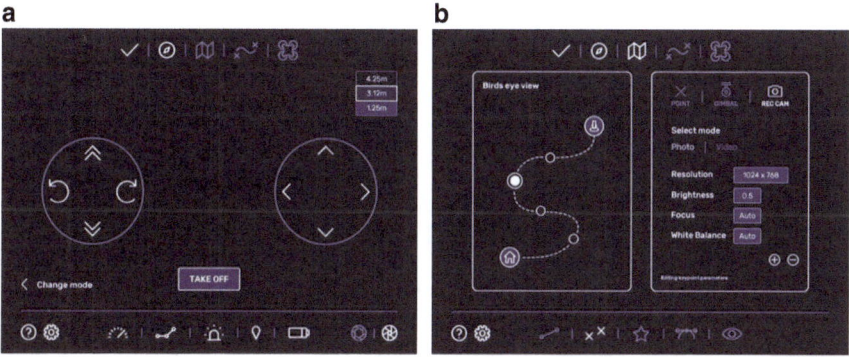

Fig. 4 (**a** and **b**) Graphical design proposal for the graphical user interface. Source: AiRT project

initiated using the Wizard of Oz Prototyping Technique (Both 2010). In the first phase, the design of paper prototypes was carried out using visual language (Fig. 2). Therefore, it was taken into account that the user interface should be able to be executed on both laptops and mobile platforms, so the models should use the standard conventions of this type of device, for example tactile gestures. The combination of cards (windows) and visual language to design the user interface of the client application represented an important change in the construction of cooperative models. On the one hand, the use of visual language made it easier to make ideas visible, tangible, and sequential and to encourage divergent collaborative thinking. On the other hand, the use of mobile cards facilitated collaborative work and improved the internal usability testing, as the repetitions were more fluid. At the end of this phase, a model of the user interface was made using the online tool NinjaMok© (NinjaMock 2018) (Fig. 3a, b). This application provided interactivity for a more realistic prototype, reproducing an interactive preview of the user interface layout and facilitating feedback from developers with the creative team and end-users. All these activities aimed to improve the application's functionality

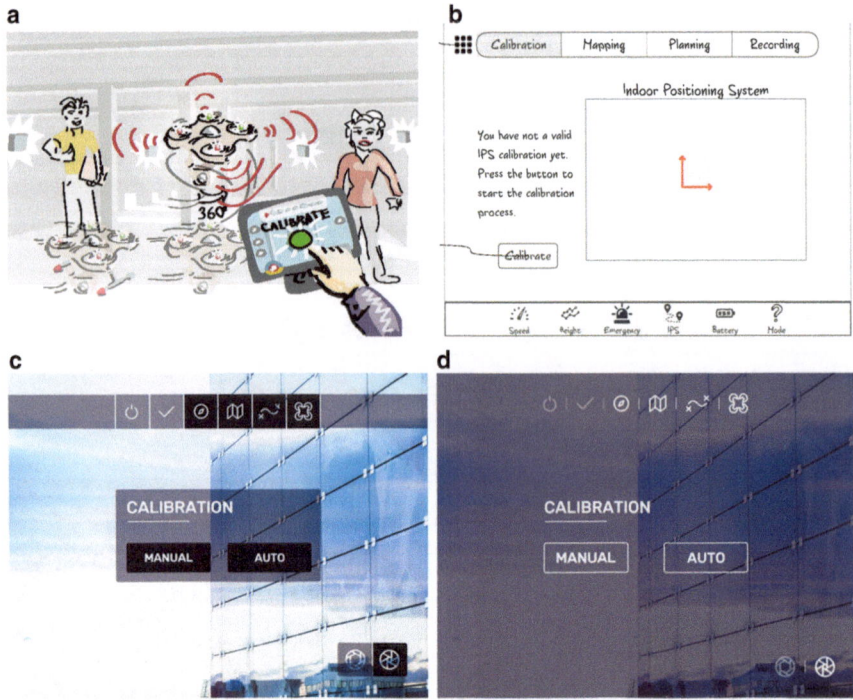

Fig. 5 (**a–d**) Example of the iterative process of graphical design of the graphical user interface from storyboard to final graphical design. Source: AiRT project

and make it more intuitive or easy to use. Once the iterative design of the model had been completed, the graphical design of the user interface (Fig. 4a, b) was developed based on the results obtained in the heuristic analysis and maintaining the iterative process (Fig. 5a, b).

From the user interface model generated in the online tool NinjaMok© and the graphical design proposal of each of the windows, a software prototype was developed that implemented the functionalities of the AiRT system. It was based on the prioritized requirements, with the aim of visualizing the solutions and identifying possible improvements.

2.4   Fifth Phase—Tests

In the last phase, end-users made use of the prototypes, based on the selection of scenarios relevant to the creative industries, in the three participating countries. The objective of this stage was to identify failures or to provide new improvements through the participation action research tool (PAR). The purpose of this technique is to obtain relevant data from experts that allow the subsequent interpretation and analysis of the facts based on the experiences (Santamarina et al. 2017). The PAR was divided into two phases. In the first phase, a user test was developed, first from the user guide and then from the prototype. In both cases, the dynamics were filmed with the aim of carrying out a subsequent analysis using qualitative data analysis software. This technique is based on the observation of the way in which a group of users carries out a series of tasks mandated by the evaluator, analysing the usability problems that they face. Finally, in the second phase, the heuristic evaluation of the AiRT system was carried out with the aim of identifying potential usability problems, checking for compliance with previously established usable design principles (heuristic principles) (Wilson 2014).

3 Analysis of the Accessibility and Compatibility of the Flight Plan Software

For the analysis of the flight plan software, 29 mesh- or mosaic-type flight plan software programs were selected for mapping and photogrammetry (Fig. 6). Of these, 10 were photogrammetry and 19 were ground station software programs. Each of them was tested for compatibility with MAVLink, APM, and PX4 communication protocols, on which platforms they could run, and to establish whether they were open source.

From the analysis carried out, it was determined that the MAVLink, APM, and PX4 communication protocols for information exchange between ground control stations and micro UAVs (unmanned aerial vehicles) were compatible with the DroneDeploy, UgCS, QGroundControl and Mission Planner, AndroPilot, APM Planner, and DroidPlanner2 software programs (Fig. 7). Regarding the compatibility with the different platforms, the software that allowed more installation options was QGroundControl, followed by Pix4D and OpenPilot (Fig. 8). On the other hand, the only open-source software programs were Opendronemap, Open MVS, QGroundControl, Mission Planner, Tower, AndroPilot, APM Planner, OpenPilot, and DroidPlanner2. In conclusion, the software that offered the best accessibility and versatility in relation to its compatibility with communication platforms and protocols was QGroundControl (Fig. 9).

4 Heuristic Evaluation of Flight Plan Software

Heuristic evaluation consists of the study and evaluation of an interface by experts, based on a set of previously defined design principles and standards. It is characterized by its fast and economic analysis, since it involves only one or several experts, who provide different answers based on the same set of rules. These standards, which serve as the basis for evaluation, are called usability principles. According to the ISO standard 9241-11, usability is defined as "the extent to which a product can

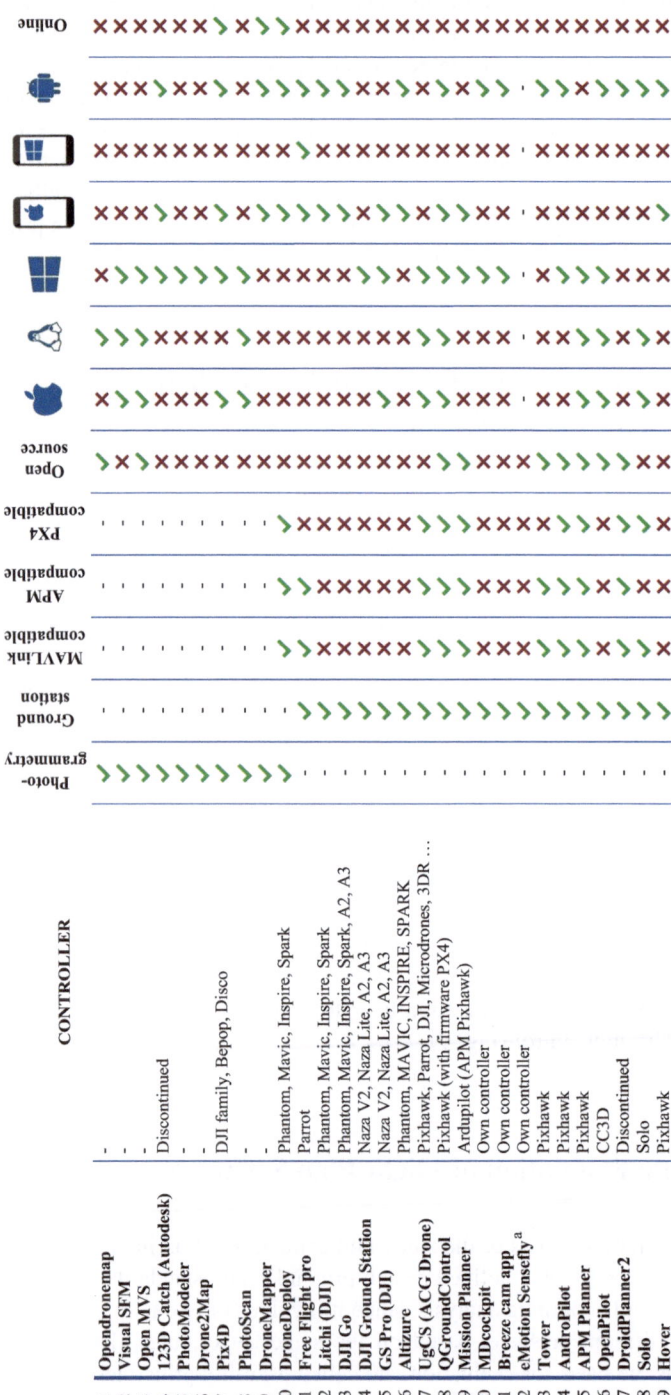

[a] All senseFly drones are supplied, by default, with eMotion flight and data management software.

Fig. 6 Mosaic or mesh flight plan software for mapping and photogrammetry. Source: Own elaboration from Intel Corporation (2015)

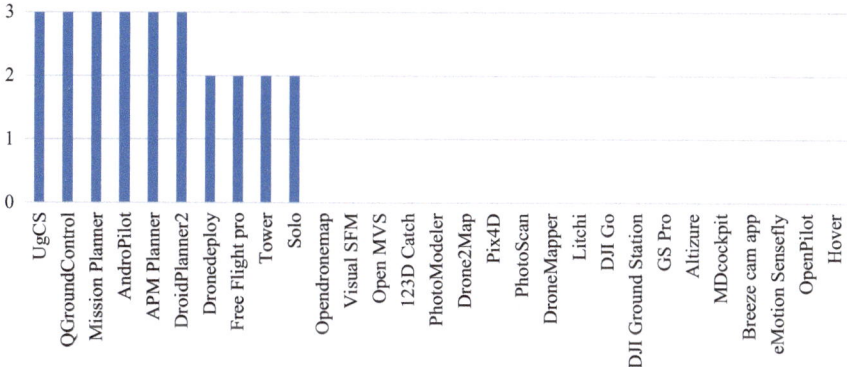

Fig. 7 Analysis of the number of compatible communication protocols (values 0–3). Source: Own elaboration

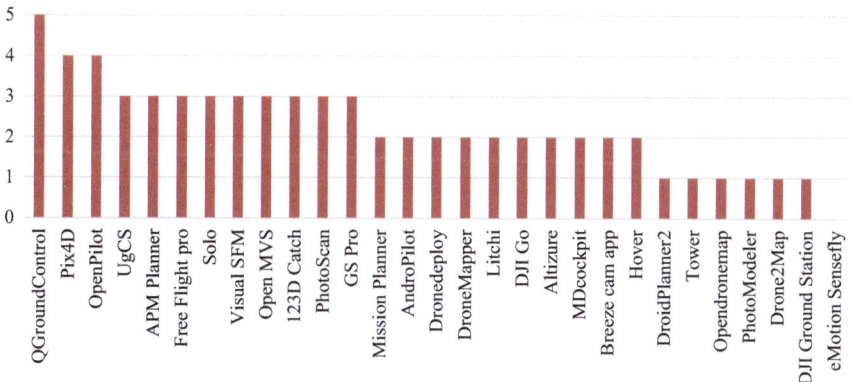

Fig. 8 Analysis of the number of compatible platforms (values 0–6). Source: Own elaboration

be used by specified users to achieve specified goals with effectiveness, efficiency, and satisfaction in a specified context of use" (ISO/IEC 1998).

The objective of this study was to identify the best designs linked to certain factors and elements of usability to avoid errors and identify opportunities for optimization in the design of the Ground Control System software and the graphical user interface of the AiRT system.

The selection of experts was carried out in compliance with the parity requirement, previous experience in the use of flight plan software, and equal participation of experts from the creative industry sector and other sectors (Fig. 10).

The experts were provided with a checklist, which contained a set of questions that assessed usability by blocks in relation to accessibility, identity, navigation, content, consistency, shortcuts, and responses to actions (Wilson 2014). Each variable had to be rated between '1', the lowest score, and '5', the highest rating. The analysis was performed through the viewing of video tutorials or the actual use

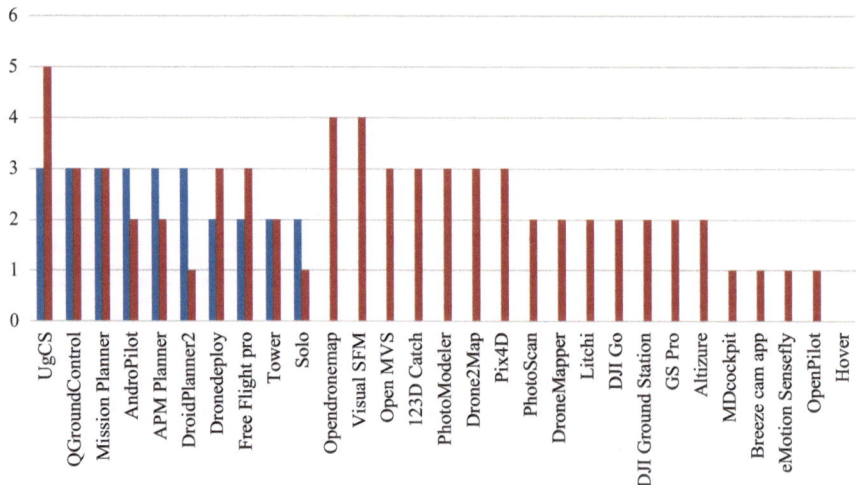

Fig. 9 Accessibility and compatibility analysis with communication platforms (values 0–6) and protocols (values 0–3). Source: Own elaboration

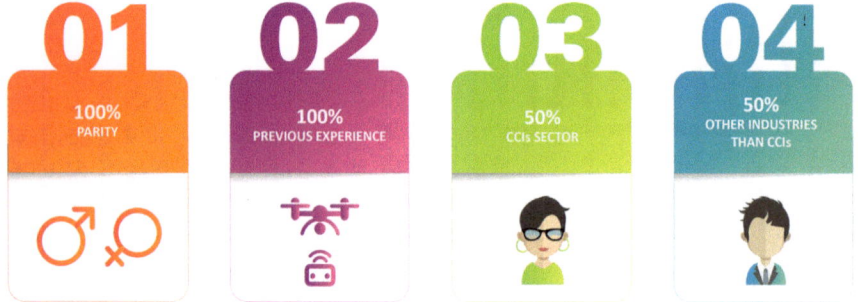

Fig. 10 Selection criteria for the experts to participate in the heuristic evaluation. Source: Own elaboration

of the applications. Only the eMotion Sensefly software, as it was linked to the purchase of the equipment (thus, it was not accessible), and the DroidPlanner2 software (discontinued) could not be analysed.

After processing the data, the highest average score linked to **accessibility** aspects was obtained by the Pix4D software with a score of 4.4 out of 5, followed by DJI Go with a score of 4.2 out of 5 (Fig. 11). In both cases, aspects related to easy software localization, downloading, and installation, compatibility with different platforms, a proper contrast between text and background, font size and spacing, and the proper use of ALT tags were highlighted.

Regarding the identity variable, the software programs that obtained the highest scores were Pix4D, DJI Go, DJI Ground Station, and GS Pro, with an average score of 4.3 out of 5 (Fig. 12). In this case, these programs stood out for the adequate

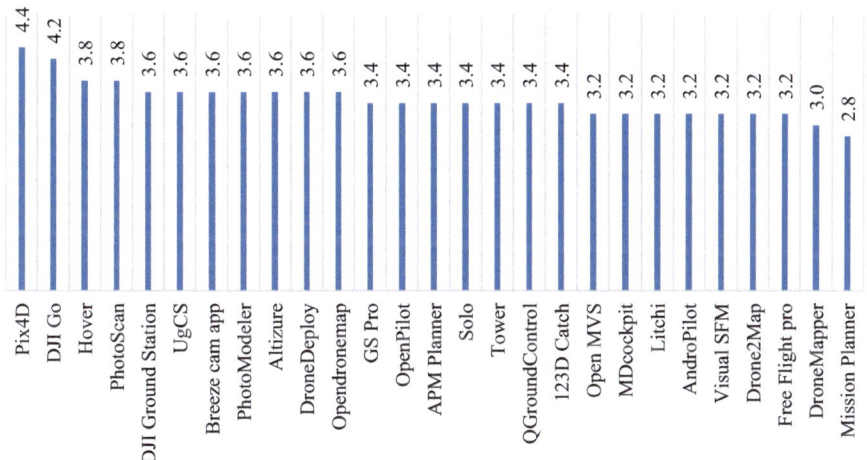

Fig. 11 Average score of the software in relation to accessibility. Source: Own elaboration

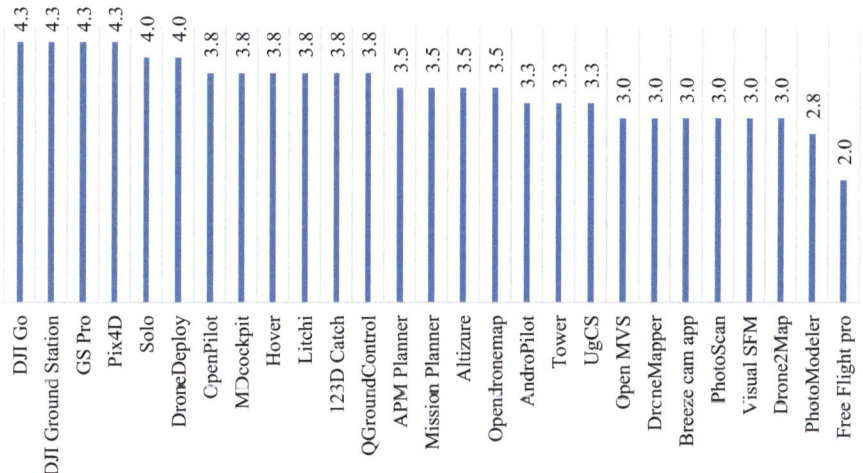

Fig. 12 Average score of the software in relation to identity. Source: Own elaboration

handling of the acronym, logo, and slogan of the software as well as for the information provided in relation to the developer company.

Concerning the navigational aspects, the software that achieved first place, with an average score of 4.4 out of 5, was DJI Go (Fig. 13). This was because it facilitates the identification, execution, and progress of tasks from the beginning, with clear and concise icons and menus, and provides adequate help support.

In content handling, the software that scored the highest was Open MVS, with a score of 4.2, followed by DJI Go and AndroPilot, with a score of 4 out of 5 (Fig. 14).

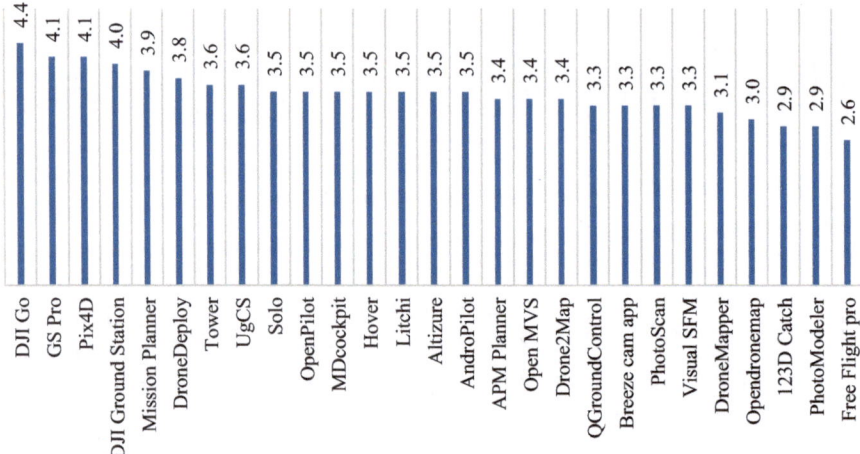

Fig. 13 Average score of the software in relation to navigation. Source: Own elaboration

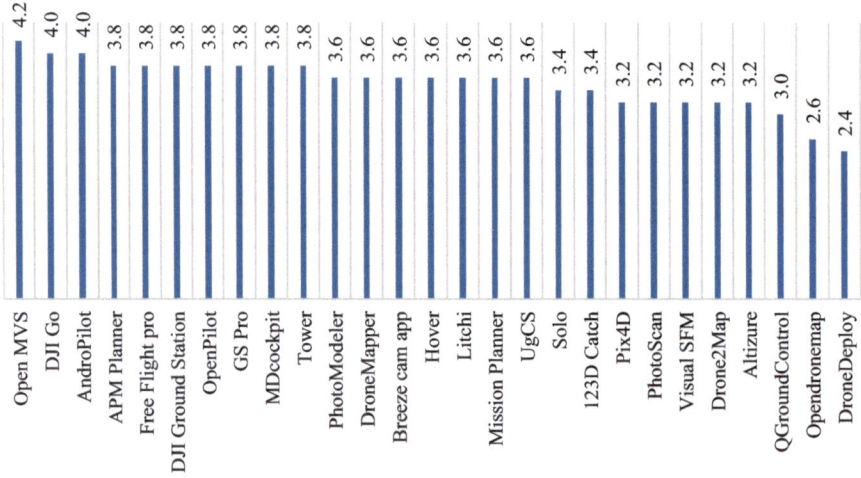

Fig. 14 Average score of the software in relation to the content. Source: Own elaboration

In all three cases, the software provided adequate handling of the main titles, with moderate use of menus and icons, and presented the critical content above the secondary content. On the other hand, the use of consistent styles, colours, and appropriate contrasts is also noteworthy.

In terms of consistency, the software with the highest rating was Open MVS, with an average score of 5 out of 5 (Fig. 15). This application presents a coherent sequence of actions in similar situations and a consistent and familiar structure of commands, screens, menus, and terminology.

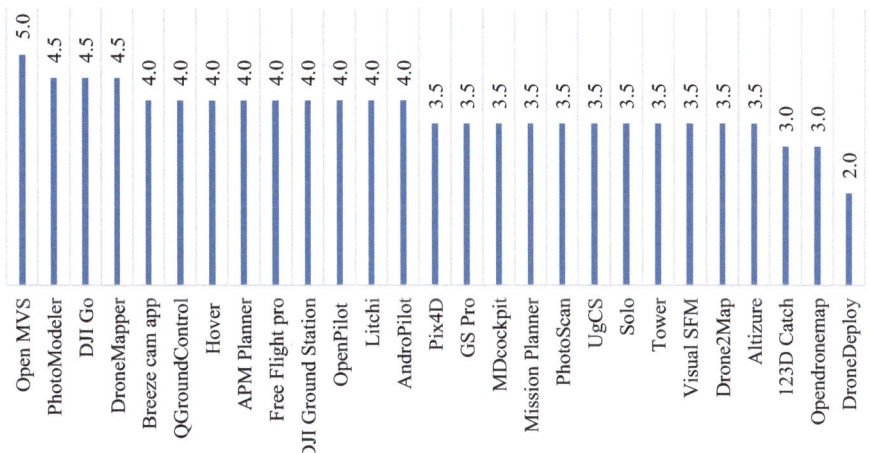

Fig. 15 Average score of the software in relation to consistency. Source: Own elaboration

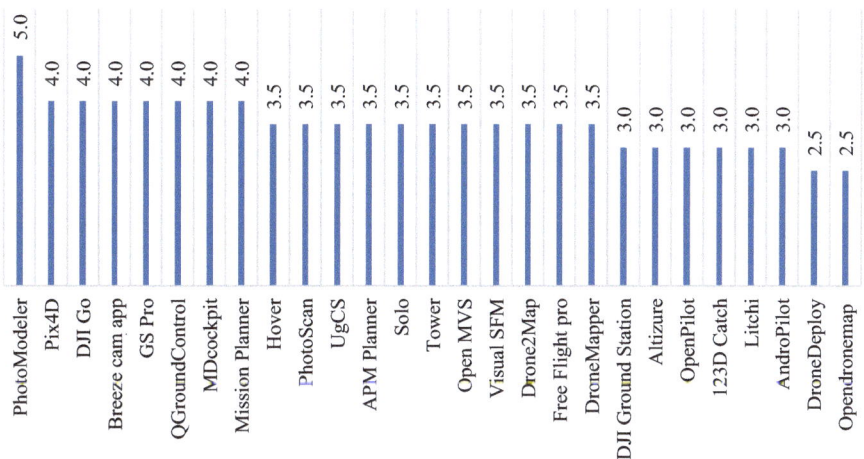

Fig. 16 Average score of the software in relation to shortcuts. Source: Own elaboration

With respect to shortcuts, the software with the highest rating was PhotoModeler, with an average score of 5 out of 5 (Fig. 16). This tool allows users to reduce the number of interactions and increase the pace of use. Moreover, it includes an action generator that enables them to customize workflows and schedule them automatically.

Finally, in relation to the response to the actions, the software programs that stood out were PhotoScan, Open MVS, Free Flight pro, and DroneMapper, with an average score of 4.3 out of 5 (Fig. 17). They presented effective time management, with effects of less than 0.1 s, and, for actions that exceeded 6 s, information on the actions is included.

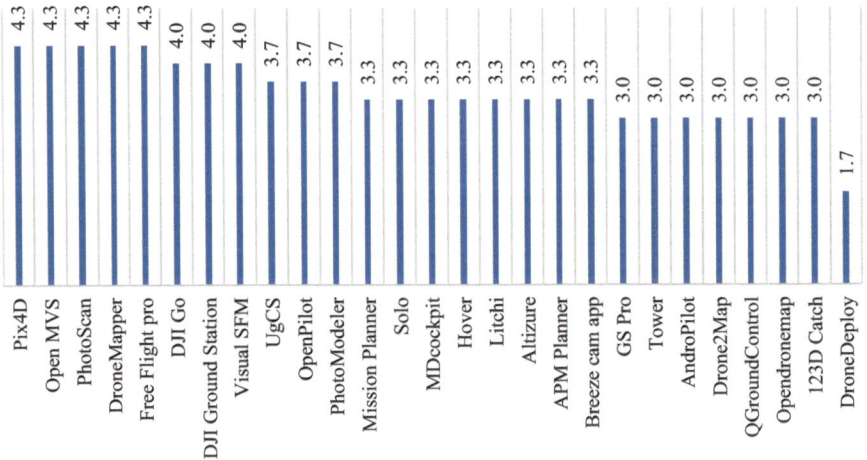

Fig. 17 Average score of the software in relation to the response to the actions. Source: Own elaboration

5 Conclusion

The combined use of the design thinking methodology together with user-centred design facilitated the construction of the Ground Control System (GCS) software and graphical user interface (GUI), taking into account the user experience.

The software analysis of flight plans facilitated the selection of the programs to be analysed and the preparation of the checklist, which was subsequently completed by the experts. The heuristic evaluation provided an approach to the best designs linked to certain factors and elements of usability of the main software available on the market. This provided the creative participants with ideas and solutions for the development of the GUI of the AiRT RPAS.

As a conclusion to the heuristic evaluation, we can report that the first positions, taking into account all the variables analysed (accessibility, identity, navigation, content, consistency, shortcuts, and responses to actions), are occupied by products of the leading brand in the market, DJI (DJI 2016) (Fig. 18). They offer intuitive, tablet-oriented applications for any type of user.

The software that was ranked in first place is DJI Go, with an average rating of 4.2 out of 5. It offers a minimalist, easy-to-use interface that displays the most basic data for easy and safe flying. The menus are well distributed, offering an excellent user experience. In the second place is the Pix4D software, with an average rating of 4 out of 5, providing a simple and clean graphical user interface that simplifies and facilitates its use. In the third and fourth places are again software from DJI, DJI Ground Station and GS Pro, with an average rating of 3.9 and 3.8 out of 5, respectively. DJI Ground Station stands out for being quite complete and offering multiple options and tools, although its aesthetic and outdated design compared with the DJI

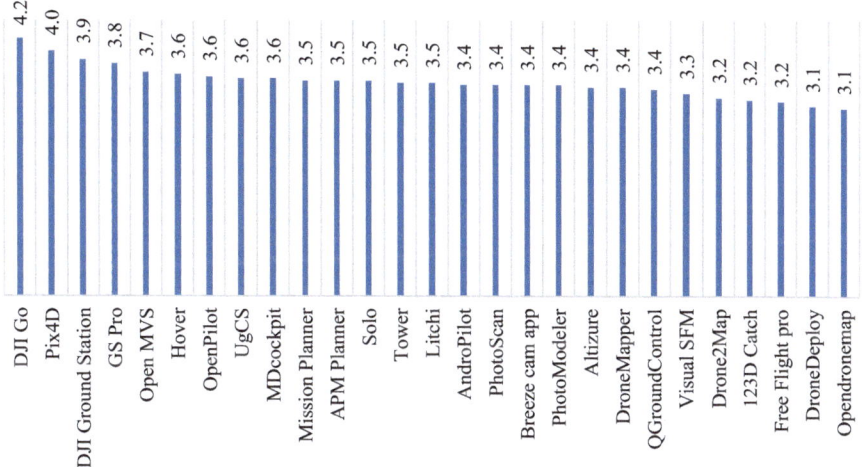

Fig. 18 Average score of the software in relation to accessibility, identity, navigation, content, consistency, shortcuts, and responses to actions. Source: Own elaboration

Go product leads to a value loss. On the other hand, GS Pro is prominent for being highly intuitive for any type of user, thus reducing the entry barrier for beginners.

References

Both T (2009) Design thinking bootleg. Institute of Design Thinking at Stanford, Stanford. Retrieved from https://dschool.stanford.edu/resources/the-bootcamp-bootleg

Both T (2010) Bootcamp bootleg. Stanford d.school. Institute of Design Thinking at Stanford, Stanford. Retrieved from http://dschool.stanford.edu/wp-content/uploads/2011/03/BootcampBootleg2010v2SLIM.pdf

DJI (2016) DJI – the future of possible. Retrieved April 21, 2018, from https://www.dji.com/es?from=store_top_nav

European Commission (2015) Research and innovation. Participant portal. Retrieved March 20, 2018, from https://ec.europa.eu/research/participants/portal/desktop/en/opportunities/h2020/topics/ict-21-2016.html

Hinton A (2004) Observing the user experience: a practitioner's guide to user research. Morgan Kaufmann. Retrieved from https://books.google.es/books/about/Observing_the_User_Experience.html?id=jIrl2L_JvZoC&redir_esc=y

Intel Corporation (2015) Ground control station | benchmark study (vol 1). Retrieved from http://dronecode.github.io/UX-Design/Research/Benchmark/GCSBenchmark.pdf

ISO (1999) ISO 13407:1999, Human-centred design processes for interactive systems. Europe. Retrieved from https://www.iso.org/standard/21197.html

ISO/IEC (1998) ISO 9241-11: ergonomic requirements for office work with visual display terminals (VDTs) – Part 11: Guidance on usability. Int Organ Stand 1998(2):28. https://doi.org/10.1038/sj.mp.4001776

Maguire M, Bevan N (2002) User requirements analysis: a review of supporting methods. Proceedings of IFIP 17th world computer congress (January), pp 231–246. doi:https://doi.org/10.1007/978-0-387-35610-5

Molich R, Nielsen J (1990) Improving a human-computer dialogue. Commun ACM 33 (3):338–348. https://doi.org/10.1145/77481.77486

Nielsen J (2012) Usability 101: introduction to usability. Nielsen Norman Group, Articles. doi: https://doi.org/10.1145/1268577.1268585

NinjaMock (2018) NinjaMock online wireframe and mockup tool. Retrieved April 21, 2018, from https://ninjamock.com/

Santamarina V, Martínez EM, Carabal MA, De Miguel M (2017) Participatory action research (PAR) in contemporary community art. In: Santamarina V, Carabal MÁ, De Miguel M, De Miguel B (eds) Conservation, tourism, and identity of contemporary community art. Apple Academic Press, New Jersey, pp 105–136

Santamarina V, De Miguel B, Segarra M, De Miguel M (2018) Importance of indoor aerial filming for creative industries (cis): looking towards the future. In: Innovative approaches to tourism and leisure. Springer, Cham, pp 51–66. doi:https://doi.org/10.1007/978-3-319-67603-6_4

Wilson C (2014) Heuristic evaluation. In: Nielsen J, Mack RL (eds) User interface inspection methods. Wiley, New York, NY, pp 1–32. doi: https://doi.org/10.1016/B978-0-12-410391-7.00001-4

How a Cutting-Edge Technology Can Benefit the Creative Industries: The Positioning System at Work

Vadim Vermeiren, Samuel Van de Velde, Michiel Boes, and Jan-Frederik Van Wijmeersch

Abstract The authors explain their innovative positioning system and its application for the indoor aerial and the creative industries. First, they analyse the current technologies available. Then, they explore the possibility of making an IPS for a creative industries RPAS by making the auto-calibration procedure robust and user-friendly. This challenge implies three parts: to create a specific hardware, to create a highly accurate multi-antenna positioning algorithm and to improve the automatic system calibration for increased user-friendliness.

1 Introduction

Remotely Piloted Aircraft Systems (RPAS) rely on accurate knowledge of their position for decision-making and control (Colomina and Molina 2014), such as:

(a) Maintaining a Stable RPAS Position By nature, without any supportive systems, the flight position of an RPAS is unstable, moving slowly out of position in a sideways direction (referred to as drifting). When this happens, the pilot is constantly forced to compensate for it by adjusting the controls in the opposite direction, which can become very tedious, and it depends greatly on the pilot's skills, potentially leading to unsafe situations. To avoid this drifting and to enable the RPAS to self-stabilise in air, the RPAS needs at least one point of reference of the surroundings. In the case of outdoor flying, GPS is used, but for indoor flying an IPS is necessary (Li et al. 2016).

(b) Guiding the RPAS to a Known Position Knowing the exact position of the RPAS within its environment, and, in addition, knowing what the environment looks like (3D model of indoor environment) enables to programme the flight path of the RPAS (Jiang and Stefanakis 2018). This is very useful, as often a camera shot needs to be repeated several times. In this case, the RPAS follows, within the IPS error, the

V. Vermeiren (✉) · S. Van de Velde · M. Boes · J.-F. Van Wijmeersch
Pozyx, Gent, Belgium
e-mail: vadim@pozyx.io

© The Author(s) 2018
V. Santamarina-Campos, M. Segarra-Oña (eds.), *Drones and the Creative Industry*,
https://doi.org/10.1007/978-3-319-95261-1_8

115

defined flight path for which repeating the scene recording does not depend on the operator's flying skills.

(c) Avoiding Collisions In a known environment with previously identified static obstacles (e.g. furniture, walls, pillars, etc.), the RPAS can swerve around and avoid them, since they are previously defined as no-fly zones.

(d) Increasing Safety While outdoors GPS can be used for safety features such as *safely landing* or *return to launch* (e.g. RPAS from *AltiGator* or *DJI*), in order to prevent a crash in the event of an unsafe situation, without an IPS and 3D indoor environment map, these safety features cannot be included for indoor use. In the case of, e.g. low battery, RPAS are commonly programmed to return to launch (starting point), and thus precise position of RPAS and starting point is needed in order to execute this command correctly.

Positioning systems can be subdivided into two groups: outdoor and indoor. Outdoor positioning systems have been well explored and standardised, using either global positioning system (GPS) or techniques that measure user position by means of a cellular network. GPS is based on the known position of determined satellites, which continuously transmit their current time and position. In contrast, cellular network positioning uses the *Global System for Mobile Communication* (GSM) to calculate position via observed time difference from two different base transceiver stations to a mobile station. For typical outdoor purposes, the accuracy of GPS (~10 m), cellular network (50–125 m) or the combination of both (~5 m) is sufficient when RPAS moves in *free space* with few but large obstacles (e.g. buildings, trees). Nevertheless, even if there would be sufficient GPS and GSM signal strength indoors, the achieved accuracy of these systems would be inadequate for indoor flights. Creative Industries need to work in spaces, where the typical height of ceilings is between 5 and 20 m, e.g. television studio set up, with many small obstacles (e.g. spotlights). Therefore cm-range accuracy is needed for RPAS use in confined spaces in order to ensure safe flights.

Over the last 10 years, indoor positioning systems have advanced greatly (Kulmer et al. 2017); however most of the developed technologies have not been specifically thought through for use with RPAS. The omnipresent need for indoor positioning in our modern way of life is reflected by the large amount of IPS developed for a wide scope of applications, e.g. medical care (e.g. location tracking of medical personnel), guiding vulnerable people (e.g. aiding visually impaired persons), emergency services (e.g. establishing emergency plans, rescue services), logistics and optimisation (e.g. accurate localisation of packages), surveying and geodesy (e.g. setting out and geometry capture of new buildings as well as for reconstructions), etc. Despite its potential for multiple applications, current IPS technologies are usually not suitable for use with an RPAS. Up to now, only two indoor positioning systems are being used for RPAS:

(a) Camera-Based Indoor Positioning (digital optical and video motion tracking system). This system consists of a large number of high speed cameras (≥250 fps) installed in the indoor environment where the RPAS will be used that will register, in

real-time, the motion of the RPAS to calculate its exact position in three dimensions. Installed cameras need to be referenced and the whole system calibrated (cameras-software-RPAS). These systems offer high precision, speed and resolution, as well as interference-free real-time tracking for engineering-related studies. Nevertheless, it should be noted that the systems were originally designed for motion capture, e.g. gait analysis when running. As a consequence, when used as IPS for RPAS, the systems have several drawbacks. These systems cost hundreds of thousands of euros, which is unaffordable for small companies, especially in the CI sector. Since the price of IPS system depends significantly on the size of the environment to control, for large installations such as film-sets, hundreds of cameras would be required, increasing the price even further. In addition, the set up of these systems is time consuming and the required infrastructure is space invading. Set up takes about 2 days and an engineer is needed for calibration of the whole system.

(b) Vision Positioning System (VPS) Unlike the camera-based IPS where the cameras are mounted in the environment, here the IPS system is located on the RPAS. The VPS mounted on an RPAS uses a combination of Ultrasonic sensors and Optical Flow Technology to control the position in environments where GPS signals cannot reach. However, VPS has several important drawbacks: the camera creates a real-time map of the ground below. It does reference the RPAS within a 3D space of a known environment (like the GPS with the help of Google Maps outdoors). Obstacles directly in the flight path are not detected and distance to home base is unknown. The system also does not work properly in low light or bright conditions (less than 100 or more than 100,000 lux). Hovering is only effective between 0.5 and 2.5 m and surfaces should not have clear patterns or texture. Transparent or reflective surfaces, or surfaces that can absorb sound waves (e.g. carpet), can even lead to severe disorientation of the system. Finally, lighting changes should be avoided in order for this system to function properly. In summary, VPS is unsuitable for typical environment conditions where creative industries might use RPAS: performing arts often have changing lightning conditions, concert halls etc. have heights over 2.5 m, filming scenarios often include carpets, bright tiles and so on.

2 The Pozyx IPS

The Pozyx IPS consists of two types of devices: anchors and tags. The tag is the device that is being tracked, while the anchors are devices with a fixed position that act as reference points for positioning. The tag can estimate its position when it knows the position of the anchors by making ultra-wideband (UWB) distance measurements to each of the anchors within range (Liu et al. 2016). This requires at least four anchors for 3D positioning, but performance is better with at least six or more anchors.

One of the requirements for a drone for the creative industries is a fast and easy deployment of the IPS system. This requires that determining the position of the anchors should be a fast process. Performing manual calibration with a laser metre is not a fast process. For competing indoor positioning systems, the deployment takes about 2 days to deploy and fine-tune the system for an area the size of a basketball court.

Because of this, we use automatic calibration that can perform the same task in seconds. The automatic calibration determines the relative positions of the anchors by making range measurements between the anchors. However, this method is very prone to measuring errors, as any error in the anchor positions will affect the final tag positioning. In addition, it does not work well if there is no clear line-of-sight and good connectivity between all anchors. Thus, a main challenge in making an IPS for a creative industries RPAS is making the auto-calibration procedure robust and user-friendly.

The main requirement for the positioning system to be usable in drone navigation is stable and accurate positioning. Wireless communication has to endure phenomena like scattering, reflections and diffraction rendering multipath propagation. Although ultra-wideband, due to its typical bandwidth of 500 MHz, is quite resistant to multipath signal distortion, the interference of multiple versions of the received signal still leads to some phase delay or even, to a lesser extent, negative phase delay in the detected rising edge of the received signal in general, UWB provides very accurate range measurements in line-of-sight (LOS) with ideal UWB antennas. In this ideal scenario, the noise behaves as almost Gaussian, with a very small standard deviation of about 3 cm. In this LOS case, a similar accuracy can be achieved for positioning. However, the indoor environment is not always ideal, with different obstacles blocking the signal which may cause non-line-of-sight (NLOS) ranging errors of up to 1 m. Because this is very dependent on the environment, it is hard to express this in terms of accuracy. It is clear that the accuracy expressed in average error will be very much dependent on how much NLOS is present, or how "challenging" the environment is. In general, metal objects, water and thick concrete walls may cause NLOS. Because of this it is better to talk about robustness. Apart from the environment, the antenna can also introduce additional errors of up to 20 cm, which are dependent on the orientation of the antenna. Both sources of errors described above (NLOS and antenna non-idealities) are hard to fix with a single antenna design. Therefore, introducing multiple antennas and thus spatial is investigated to significantly reduce the error on the positioning quality in challenging environments.

3   An RPAS IPS for the Creative Industries

3.1   Creation of Specific Hardware to Be Mounted on Drones

Figure 1 shows the adapted high-level overview of the Pozyx IPS for drones. It has the following features:

It consists of a central controller unit, equipped with an Inertial Measurement Unit (IMU) and an altimeter, connected to four UWB units to be mounted on the corners of the drone. DecaWave DW1000 chips are used for UWB communication.

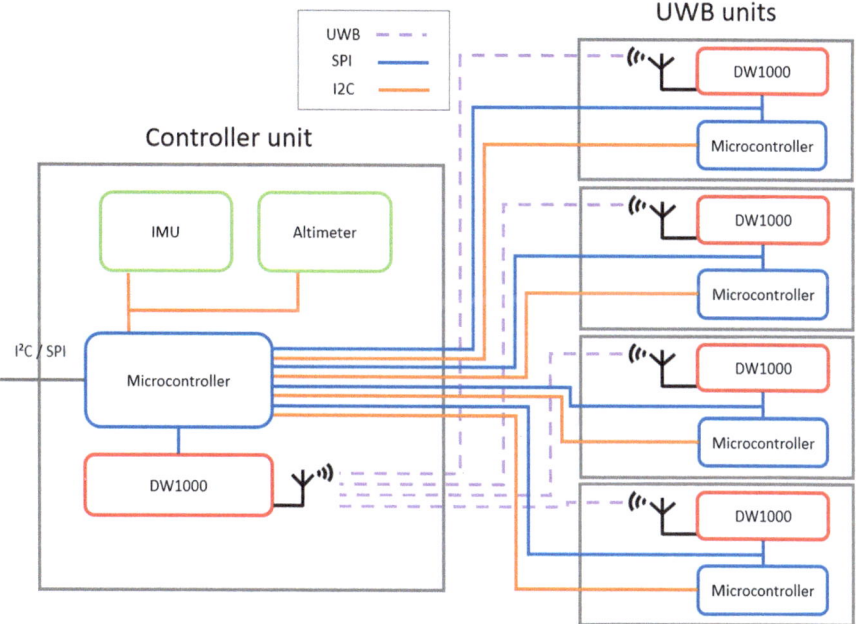

Fig. 1 Adapted high-level overview of the system on the drone. Source: own elaboration

Both the controller unit and the UWB units include a microcontroller. This design offers the versatility to either work with the controller module as master and the UWB units as slaves (with the possibility of outsourcing part of the computation load to the slaves) or to operate the UWB units as stand-alone tags.

Again, in order to offer versatility, three separate ways of communicating between the controller and UWB units are supported. One can use the UWB signal (decaWave chip to decaWave chip), the SPI bus (central microcontroller to slave decaWave chips) or the I2C (central microcontroller to slave microcontrollers).

The central controller unit is, in essence, a standard Pozyx tag, while the UWB modules are custom designed for use on RPAS. The design for these can be seen in Fig. 2 [UWB module design (right) and device (left)], which includes the decaWave chip on the left-hand side, a microcontroller on the right-hand side, an integrated antenna on the top and both an SPI and an I2C interface at the bottom.

3.2 Creation of Highly Accurate Multi-antenna Positioning Algorithm

Initially, each UWB module is positioned independently, resulting in four independent positions. Subsequently, the known configuration of antennas on the drone is mapped onto these measured positions. For this mapping, Procrustes analysis is

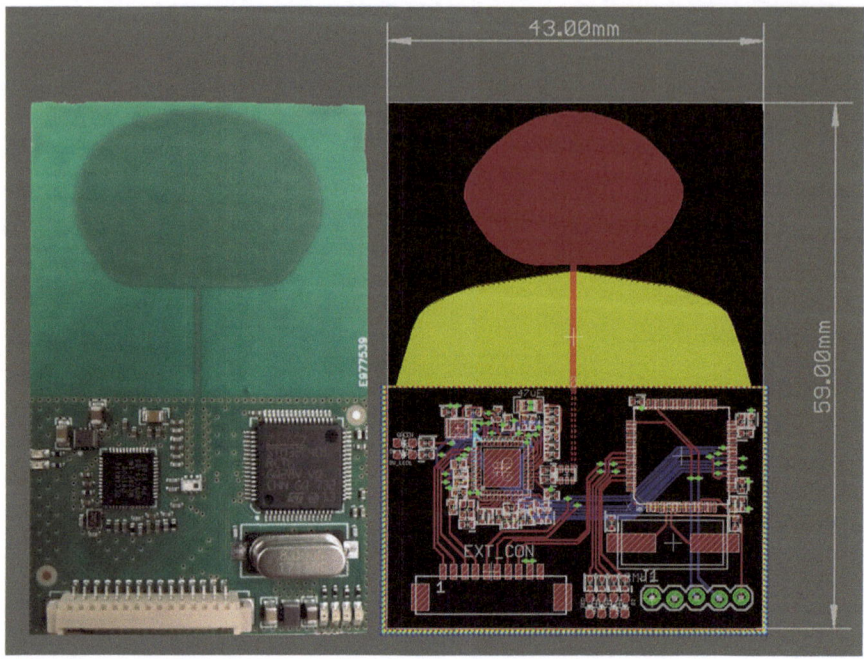

Fig. 2 UWB module design (right) and device (left). Source: own elaboration

used, without scaling, as the size of the drone is fixed. The translation component of the Procrustes analysis results in the position of the centre module of the tag, which, in case of default placement of the UWB modules on the corners of the drone, is the average of the antenna positions: $x_{tc} = \frac{1}{4} \sum_{i=1}^{4} x_{ti}$, and similar for the y- and z-coordinates.

The rotation component is calculated algebraically as follows. First A is defined as a matrix consisting of the coordinates in the drone coordinate system of antenna ti in column i of the matrix, and B as consisting of the coordinates in the anchor coordinate system, after translation to the origin as described above, of antenna ti in column i. The rotation component is then found as the rotation matrix R for which $\|RA - B\|$ reaches its minimal value. This is known as the Orthogonal Procrustes problem in algebra, and its solution is given by performing a singular value decomposition of $BA^T = M = U\Sigma V^T$, after which R can be found as $R = UV^T$. The Euler angles of the drone can subsequently be deduced from its rotation matrix R.

This method is computationally relatively light and effectively reduces the random noise on the positioning as $\sigma_{tc} = \frac{1}{4} \sqrt{\sum_{i=1}^{4} \sigma_{ti}^2} = \frac{1}{2} \sigma_t$, as all σ_{ti} are considered the same, equal to σ_t. Thus, this approach halves the random noise on the positioning, as compared to the single antenna NLLS method.

On the other hand, this method is relatively sensitive to outliers and failures in the measurements. An outlier in one of the range measurements will result in an outlier in one of the antenna positions, if it is not detected and rejected. And in turn, an outlier in one of the antenna positions, if undetected, will cause an outlier in the calculated drone position and orientation. Thus, outlier detection is necessary on both the antenna positioning and drone positioning level. This can be achieved by calculating the a posteriori aberration of each of the measurements, subsequently rejecting aberrations larger than a set threshold, and recalculating the position based on only non-rejected measurements. For the antenna positioning, this implies that first the position is calculated using all measurements. Next, the measured distances to the anchors are compared to the distances based on the calculated antenna position. If the difference between calculated and measured position is above a predetermined threshold, the measurement is considered an outlier, and finally the position is recalculated without the outlier distance measurements. This procedure can be repeated until no more outlier measurements are detected, or the number of valid anchors becomes too low for reliable positioning. The method for outlier rejection for the drone positioning is fully analogue.

Although this is an effective technique for outlier rejection, the hierarchical method of first positioning all antenna's and then positioning the drone based on the antenna positions implies that useful measurements can also be discarded. Indeed, when an antenna has insufficient valid range measurements for reliable positioning, its position cannot be calculated, and also the valid measurements are discarded. Thus, in some cases, this method will not be able to position the drone, even though the overall number of valid range measurements would be sufficient, just because valid range measurements were discarded on the antenna positioning level. In practice, on the other hand, this issue does not cause any problems in normal environments.

For validation, a comparison with GPS was made in an outdoor environment on a scale of 20 m × 20 m. The results are shown in Fig. 3. From this, it can be seen that both systems perform good positioning. However, the GPS and IPS do not completely match. Due to the lack of accurate ground-truth, it cannot be determined if this is due to the IPS or the GPS. However, it is known that the GPS is not as accurate and that it relies more on smoothing to obtain a smooth trajectory.

Additionally, drone positioning measurements were made with the drone on three key locations within the convex hull formed by the anchor positions: one at a corner, one at an edge and one in the centre. The resulting CDFs of the positioning error are shown in Fig. 4. Even in case of the corner position, in which the worst positioning quality is expected, the vast majority of errors fall below 25 mm, and occasional positioning outliers still fall below 200 mm. The other locations perform even better. Combined with the comparison with GPS data, this measurement allows us to conclude the IPS is performing as expected.

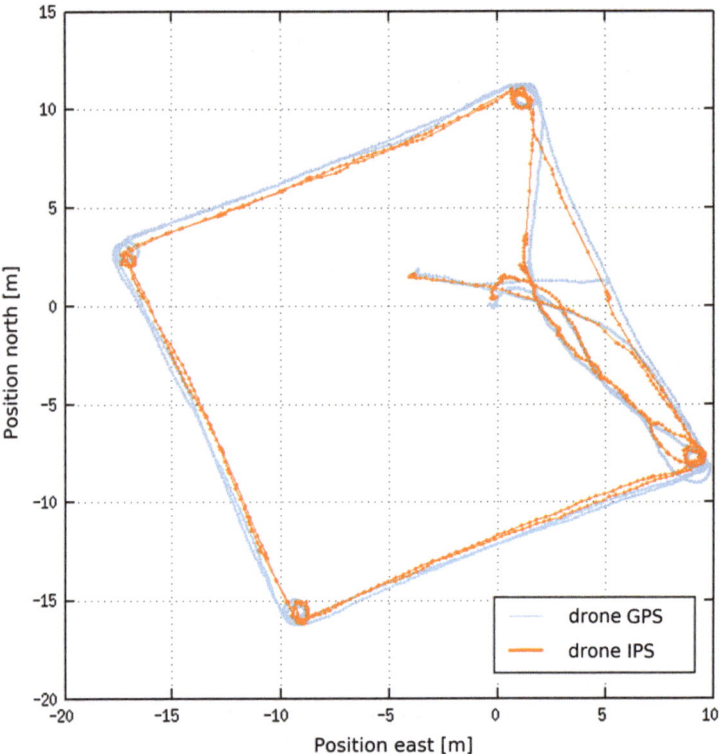

Fig. 3 A comparison in positioning between a drone GPS and the developed IPS. Source: own elaboration

3.3 Improvement of Automatic System Calibration for Increased User-Friendliness

In its basic form, automatic calibration relies on all inter-anchor distances to calculate the anchors' relative positions to each other, as shown in an example setup in Fig. 5. Assume anchor a is on position $[x_a, y_a, z_a]$, and the distance between anchor a and b is measured to be $\widehat{d}_{ab}$. The goal of the auto-calibration procedure is to find $[x_a, y_a, z_a]$ for all a, based on $\widehat{d}_{ab}$ for all pairs of anchors a and b that are within range of each other. This can be achieved using a non-linear least squares (NLLS) procedure with cost function defined as follows:

$$f_{cost}(x_a, y_a, z_a, \ldots) = \sum_{a, b} \left(\widehat{d}_{ab} - \sqrt{(x_a - x_b)^2 + (y_a - y_b)^2 + (z_a - z_b)^2} \right)^2$$

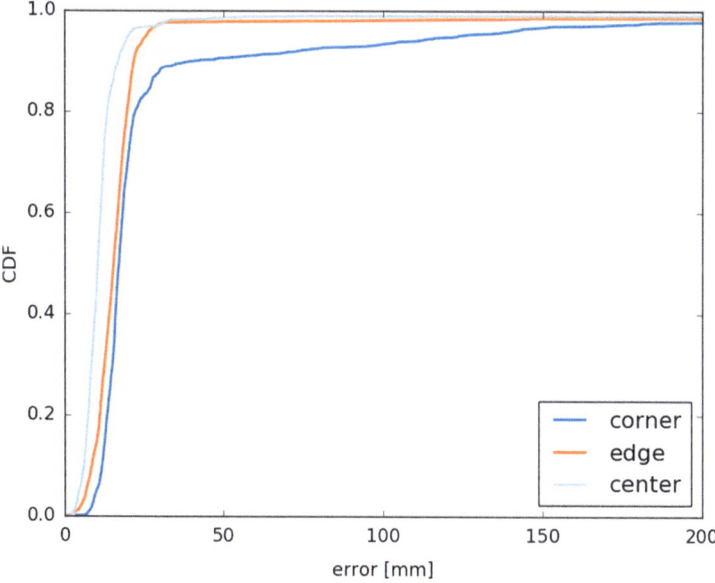

Fig. 4 CDF of drone positioning error for three different key locations in the convex hull of anchors. Source: own elaboration

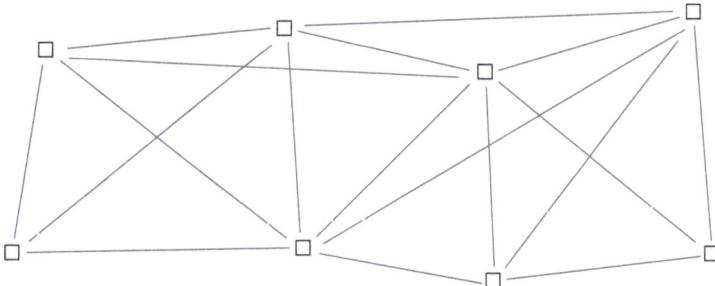

Fig. 5 Example setup in which anchors are visualised as squares. The lines connecting them represent the measured inter-anchor distances, to be used in auto-calibration. Source: own elaboration

As this is a minimisation in a highly dimensional space ($3N_a$ for N_a the number of anchors in the system), this is a relatively computationally heavy operation. Unlike positioning though, this needs to be executed only one time during initialisation of the system.

In most practical setups, the above-described basic auto-calibration will have trouble finding an optimal configuration, and additional constraints are necessary to make the minimum search converge to a solution. Normally, the assumption is made that all anchors are on the same height, $z_a = h, \forall a$, as this is the case for the majority

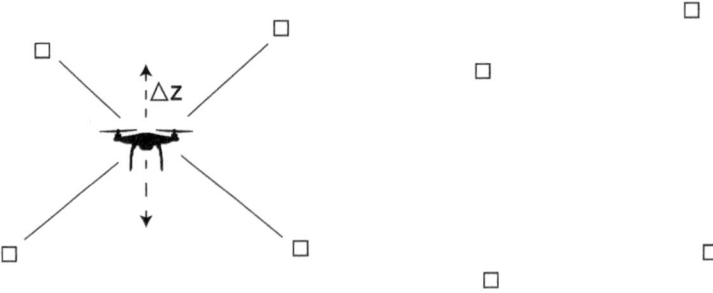

Fig. 6 Example setup in which anchors are visualised as squares. The additional distance measurements between tag and anchors during a short reference flight, varying the z value of the tag, are visualised as lines. Source: own elaboration

of setups. If this is not the case, the heights of the majority of the anchors will need to be measured manually in order to obtain correct results from the basic auto-calibration algorithm.

An alternative could be Simultaneous Localisation and Mapping (SLAM), which is a technique that is often used in robotics and navigation. In the context of this project, it would involve calculating the coordinates of the anchors while positioning the tag at the same time, based on distance measurements between the different tag antennas and the anchors they have within range. Unfortunately, this technique is both computationally very heavy and not very accurate. In addition, it does not take the measured distances between the anchors into account.

Thus, a method was developed in which not only the inter-anchor ranges are used (like in the basic auto-calibration algorithm), but also measured ranges between the tag antennas and the anchors (like in SLAM), as visualised in Fig. 6. Like in the basic auto-calibration algorithm, and unlike in SLAM, the calibration is calculated after all measurements are done.

As the tag has an altimeter on board and it can keep track of its relative height by comparing altimeter data at the times of ranges measurements to the anchors within range. In case the drone's calibration flight is initiated at zero height, this relative height can be seen as an absolute height measurement by comparing to the altimeter measurement at the beginning of the calibration flight. This will be a valid z-axis value, as long as the calibration is sufficiently short, compared to random drift on the altimeter. Thus, the auto-calibration method as described in the previous section can now be run with an increased amount of "virtual anchors", which are the drone antennas on different places during the calibration flight, of which the z value is known. For smaller setups, in which a short calibration flight suffices to make range measurements from the tag to all anchors, this removes the need to manually measure z values of all anchors, as they can now be deduced from the known z values of the "virtual anchors".

For larger setups, in which a short calibration flight does not suffice to make range measurements to all anchors in the setup, it can no longer be assumed that the

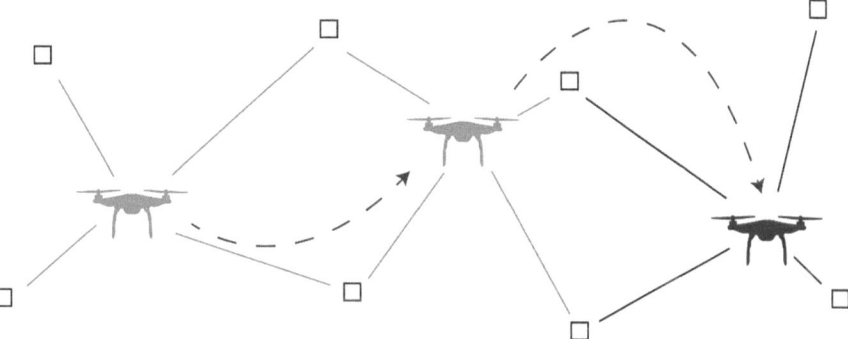

Fig. 7 Example setup in which anchors are visualised as squares. A selection of additional distance measurements between tag and anchors during the reference flight are visualised as lines. Source: own elaboration

z-coordinates of the "virtual anchors" are still known, because of random drift on the pressure sensor data. Differences in z-coordinates between two consecutive ranging measurements from the tag antennas to the anchors are still known though, as the time difference between these is rather limited. Thus, instead of assuming the z-coordinates of the "virtual anchors" to be known in the auto-calibration cost function, extra terms can be introduced in the cost function to capture the known height differences between them:

$$f_{cost,z} = \sum_i \left((z_i - z_{i-1}) - \Delta z_{i,i-1} \right)^2$$

where z_i is the height of "virtual anchor" i, being the tag on the ith consecutive measurement, and $\Delta z_{i,\ i-1}$ the height difference between the ith and $i-1$st measurement, as measured by the pressure sensor. The total cost function to be optimised for full auto-calibration is then $f_{full} = f_{cost} + f_{cost,\ z}$, with the additional knowledge that, in the assumption of a calibration flight that starts and ends on the ground, $z_0 = z_N = 0$, with N the number of distance measurements done during the calibration flight. In Fig. 7, an example setup with a long reference flight, including some extra distance measurements is shown.

In order to determine the performance of the auto-calibration. The default auto-calibration process was repeated several times and the variation of the outcome was logged. In Fig. 8, the cumulative distribution function is shown for the anchor positioning errors. In total, the calibration process was repeated 30 times in a static environment with four anchors in an area of 100 m². The experiment was performed in two scenarios: one scenario with all anchors in LOS of each other, and the other scenario with one anchor behind a wall (NLOS). It can be seen that the errors for this experiment are very low, between 10 and 30 mm, even in a partial NLOS-environment. The reason for this level of accuracy is due to the abundance of measurements that are used for the auto-calibration. Each inter-anchor measurement is repeated 50 times, which significantly reduces the error. These additional

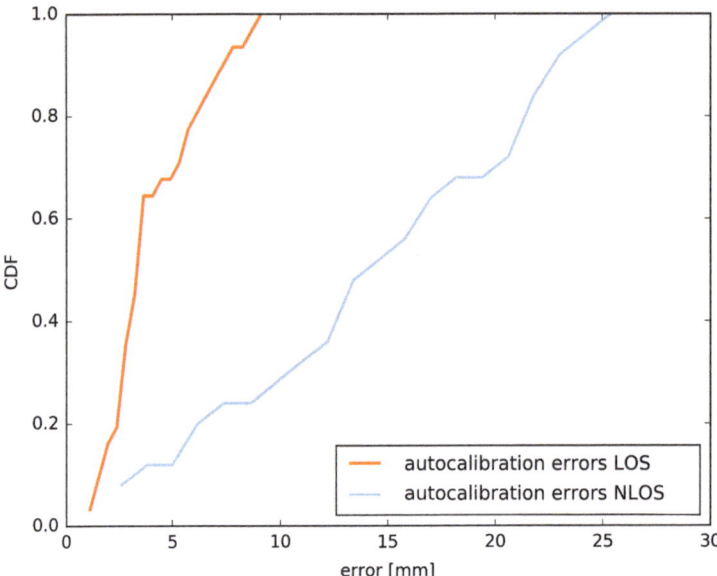

Fig. 8 CDF of anchor positioning errors after auto-calibration. Source: own elaboration

measurements take extra time; however, because the calibration process must only be performed once, it is of little importance. In total, the anchor calibration process takes no longer than 5 s.

4   Conclusion

In this chapter, we have explained Pozyx's positioning system and its application for the indoor aerial and the creative industries. We have demonstrated that UWB-based indoor positioning systems are suitable for drone integration and provide sufficient precision to allow professional high-quality filming. After analysing the current technologies available, we have explained the possibility of making an IPS for a creative industries RPAS by making the auto-calibration procedure robust and user-friendly. This challenge has been developed in three steps: creating a specific hardware, generating a highly accurate multi-antenna positioning algorithm and improving the automatic system calibration for increased user-friendliness.

References

Colomina I, Molina P (2014) Unmanned aerial systems for photogrammetry and remote sensing: a review. ISPRS J Photogramm Remote Sens 92:79–97

Jiang W, Stefanakis E (2018) What3Words geocoding extensions. J Geovis Spat Anal 2(1):7. https://doi.org/10.1007/s41651-018-0014-x

Kulmer J, Hinteregger S, Großwindhager B, Rath M, Bakr MS, Leitinger E, Witrisal K (2017) Using DecaWave UWB transceivers for high-accuracy multipath-assisted indoor positioning. In: Jamalipour A, Papadias C (eds) 2017 I.E. international conference on communications workshops (ICC Workshops). IEEE, Paris, pp 1239–1245

Li B, Zhao K, Saydam S, Rizos C, Wang J, Wang Q (2016) Third generation positioning system for underground mine environments: an update on progress. http://www.ignss2016.unsw.edu.au/sites/ignss2016/files/u80/Papers/non-reviewed/IGNSS2016_paper_28.pdf. Accessed 16 Mar 2018

Liu R, Yuen C, Do TN, Guo W, Liu X, Tan UX (2016) Relative positioning by fusing signal strength and range information in a probabilistic framework. In: Paper presented at the 2016 I.E. Globecom Workshops (GC Wkshps), Washington, DC, 4–8 Dec 2016

Indoor Drones for the Creative Industries: Distinctive Features/Opportunities in Safety Navigation

José-Luis Poza-Luján, Juan-Luis Posadas-Yagüe, Alberto Cristóbal, and Miguel Rosa

Abstract This chapter provides an analysis of the indoor drones' characteristics and differences from the existing RPAS with a special focus on the features that indoor drones must have. The authors offer a review of the drone's characteristics in order to propose a vision about the main functionalities that a drone can perform. The work provides a vision about the components that integrate a necessary architecture to a safety flight in indoor environments.

1 Introduction

Creative industries require the ability to control cameras frequently. Many devices such as cranes, rails or portable frames are used in order to obtain interesting shots. These devices often have many drawbacks: they are complex to install, handle and remove, the devices have limited movement space and they are invasive on the scene they are recording. Devices like Steadicam may avoid these drawbacks, but they cannot be used in any situation and they cannot be moved off of the ground.

On the other hand, unmanned aerial vehicles (UAV) or drones, also known as Remotely Piloted Aircraft System (RPAS), which consider as the whole system (including the ground control), obviate the drawbacks. When recording takes place indoors, let's say a television or movie set, drones can provide shots that are not available to current auxiliary devices because of their stability and precision (Castillo et al. 2007).

Drone navigation requires knowledge of the position of the drone at all times. In outdoor flights, drones can use GPS location systems. When working indoors, GPS does not have the accuracy to allow for a safe flight, and therefore an Indoor

J.-L. Poza-Luján (✉) · J.-L. Posadas-Yagüe
Instituto Universitario de Automática e Informática Industrial (ai2), Universitat Politècnica de València, Valencia, Spain
e-mail: jopolu@upv.es

A. Cristóbal · M. Rosa
AeroTools-UAV, Madrid, Spain

© The Author(s) 2018
V. Santamarina-Campos, M. Segarra-Oña (eds.), *Drones and the Creative Industry*,
https://doi.org/10.1007/978-3-319-95261-1_9

129

Fig. 1 Main areas to be
considered in the design of
an autonomous indoor
drone. Source: own
elaboration

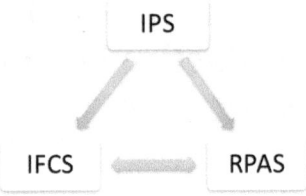

Positioning System (IPS) is needed. Furthermore, due to smaller spaces and increased risk of damages to property and people in case of an accident, much higher accuracy is required. Typically, the necessary accuracy is in the order of tens of centimetres, which means two orders of magnitude more precise than GPS.

If the drone incorporates an internal positioning system, it is possible to control it remotely and autonomously. This control must provide the mission to the drone, monitor said mission and be able to ensure all safety features. To perform all these tasks, it is necessary to have an Intelligent Flight Control System (IFCS). This system must be integrated with the IPS, since it requires knowledge of the position to generate a map of the interior space. It must also be integrated with the drone, since it must be able to control it with guarantees. Consequently, a drone mission will consist of a sequence of actions that the drone must perform. Therefore, the drone must be equipped with different flight modes. These modes are one of the strengths of the chapter, since it is the key to security (Fig. 1).

Currently, indoor drone navigation is mostly performed using commercial, off-the-shelf solutions, both for drone control (Hussein et al. 2015) and for trajectory tracking (Santana et al. 2014). This latter aspect is one of the most interesting as far as research is concerned (Martínez and Tomás-Rodríguez 2014).

The drone's real-time navigation and control features are especially relevant to the creative industry. Safety is a key factor for both outdoor and indoor flight environments. In most countries, outdoor drone flights in populated areas are very restricted. Common outdoor flight environments are in non-populated areas with few elements of value around. However, even if indoor flights are not regulated, it is not strange to find elements of value like paintings, sculptures, lamps, furniture and so on. Combined with smaller spaces and the presence of people, indoor drone flights should have higher levels of security than outdoor flights. This aspect therefore determines, to a large extent, all questions relating to the design of both the control architecture and the drone.

2   Drone Characteristics

The core of the indoor system is the drone. Drone must have the necessary components to ensure the safety requirements. In this section, we describe the drone classification and the review of the components that are directly related with the drone.

2.1 Drone Classification

A classification of drones in their different form-types and configurations can be carried out based on different approaches. A very frequent classification takes into account the ways the drone gets lift and, separately, the way the drone takes off. A basic scheme could be formed by drones in the following configurations: Multi-rotor, Fixed-Wing and VTOL (Vertical Take-off and landing).

Each of these configurations has pros and cons and specific applications where it is the most suitable option. Since the scope of this book is on how technology is applied or used in indoor drones, it is worth focusing on the type of drones which are more oriented to be used in limited airspace.

These types of drones are the multi-rotor or multi-copter types, which have a list of advantages that can be summarised as follows.

- Vertical take-off and landing, which minimise required space on land for operation. This advantage is reduced if compared with the small and lightweight Fixed-Wing, which can be launched by hand or using small catapults, while the landing can be done with parachute or controlled loss of lift.
- Possibility of stationary or very low speed flights.
- Better manoeuvrability and accuracy while flying. Fixed-wing systems fly with wide curvilinear trajectories, on a high turning radius, as well as both restricted ascent and descend speeds. In contrast, multi-rotors can fly following any trajectory on a 3D path, allowing a better approach to the target.
- Due to configuration and design, payloads are generally heavier and bigger related to the aircraft's size.

The multi-rotors can be classified also by number of motors:

- Bi-copter: two motors. They need adjustable pitch on motors and propellers in order to fly in a balanced manner.
- Tri-copter: three arms with a motor at the end. The tail motor must have a mechanical system to counteract the torque generated by the other two motors.
- Quad-copters: four arms with a motor at the end.
- Hexa-copter: six arms and six motors.
- Octocopter: eight motors either on eight arms or four arms (biaxial configuration).
- There are multi-rotors with 10, 12 and up to 18 motors, most of them just in trial mode having not reaching commercial status.

When used specifically for indoor environment, there are a number of restrictions that must be addressed and sorted out with new developments of the technology. This is the case of positioning and navigation in airspace where the main source of reference for positioning, the well-known GPS (Global Positioning System) or in general terms, the GNSS (Global Navigation Satellite System) devices, have limited coverage and the signal is not reliable.

Nowadays, almost every positioning device in the RPAS market works GNSS mode, which allows the receiving of data from different satellite constellations (GPS,

Glonass, etc.). They can also receive WAAS (Wide Area Augmentation System) or EGNOS (European Geostationary Navigation Overlay System) differential corrections, both of which are based on SBAS (Satellite Based Augmentation System), which provides differential positioning that increases accuracy of the GPS signal in almost every case.

This system can also fix typical errors in GPS signals, such as clock status, ionosphere influence or errors in the orbital tracks. Using this system, positioning accuracy is increased from 2–5 m to 1 m. This 3DGPS precision is enough to navigate and provide geo-referenced images.

RTK systems (Real Time Kinematic) improve positioning and navigation, reaching an accuracy of a centimetre. The main factors to be considered when selecting RTK GNSS systems are:

- Concurrence
- Frequencies (L1 and/or L2 in the GPS case)
- Time to "fix" the signal

This system needs its own referencing station or a supplier of differential corrections. These corrections are transmitted via radio or cellular net signals. The GNSS receiver used as a reference shall have, at the very least, same features and performance as the mobile receiver, meaning that if the mobile receiver is a double frequency GNSS receiver (L1 & L2), the base receiver shall be also a double frequency GNSS device, or the whole system will see its performance reduced.

The starting sequence (automatic) is the process through which the receiver solves ambiguities and goes from autonomous accuracy to centimetre precision. In L1 frequency devices, it takes about 20 minutes, an aspect that must be considered for the operation, as it can cause delays when signal is lost and it has to be reinitiated. Dual frequency devices are faster in this process.

With regard to indoor positioning, where the GPS signal is not available, other systems must be considered. One that should be highlighted is the UWB (Ultra-Wide Band) Technology that uses radio-frequency signal in the wide band spectrum to measure distance to beacons and deduct positioning by triangulation. This type of signal can penetrate walls and provides more accuracy than others like GPS, Wi-Fi (IEEE 802.11), and so on.

Bluetooth systems can be an alternative, although accuracy is very low and is usually discarded for most of the applications

2.2 Navigation Systems

A drone mission is a set of actions that a drone must do in a specific environment. The actions, for example, are to take a picture, start a video recording, go to another point in the space, change the pose of the drone, and similar actions. All these actions must be controlled. Consequently, the navigation system in a drone is necessary to control the drone mission.

2.2.1 Inertial Measurement Unit (IMU)

Gyroscope

There a number of gyroscopes that make use of different physical phenomena to obtain a measure of the rotation speed (MEMS, gyro-laser, mechanical, etc.), although the most used in RPAS is of the MEMS type, due to its advantages in size, weight, and also important, in price (as it is used in the mobile phone devices market).

The MEMS-type gyroscopes keep a mass in constant vibration and periodically measure the deviation with regard to the initial plane of vibration. The deviations allow the obtaining of a measure of the Coriolis Force experienced by the mass, which is originated by the rotation speed of the gyroscope.

Accelerometer

In the case of accelerometers, as it is with the case of the gyroscopes, the main type is the MEMS type. The way they function is very similar. The inner part of this sensor is composed of a comb type structure that forms a series of capacitors. The mobile electrodes are fixed to a known mass that is also connected to a device that will return the mass to a known position. When the sensor detects acceleration, the mass moves and provokes a change in the capacitance of the capacitor. Measuring this capacitance allows estimation of the acceleration experienced by the sensor.

Magnetometers

Also, most used magnetometers are of the MEMS type. These magnetometers are based on the Hall effect to function. This effect is due to the difference in potential that appears when the forces in a conductor are not distributed evenly. By measuring this difference in potential, the magnetic field in the sensor can be estimated, along with the heading with regard to the earth's magnetic field.

2.2.2 Barometer

Most barometers used in RPAS are piezoelectric barometers (based on MEMS type). A piezoelectric element is fixed at the only exit of a cavity, completely sealing it. In this cavity, an air mass is trapped with the reference pressure, generally 10,1325 Pa. The pressure differences between the air in the cavity and the atmospheric air provoke forces that act on the piezoelectric element, causing a deflection and then a difference in electric potential that can be measured. Based on this difference in potential and correcting with the sensor's temperature, the atmospheric pressure can be deducted.

2.2.3   Ultrasounds

Ultrasound sensors are based on the same principles involved in sonar. They send a sound wave at a very high frequency and measure the time that the echo takes to reach the sensor again. This time, by multiplying by the speed of sound and dividing by 2, the sensor can calculate the distance from surrounding obstacles. In RPAS, they are generally used as altimeters (near the ground) or to avoid large-sized obstacles such as walls or ceilings.

2.2.4   Infrared

The use of infrared sensors is very restricted for outdoor operations due to a lack of sunlight, so they are limited indoors for measuring very short distances (when landing) or distances to very big obstacles (walls or ceilings).

2.2.5   Optical Flow

The optical flow cameras allow measuring of ground speed of an RPAS using images of the land the RPAS is flying over. The way they function is very similar to a current typical computer mouse. A low-resolution camera (lower than 1 MP) takes pictures at a very high refreshing rate and feeds an algorithm that can extract patterns from the images. Changes in the pattern between two consecutive images and time for the changes can give the horizontal speed of the RPAS, and the new position starting from a known initial position. In principle, any camera with a relevant refreshing rate is valid for this purpose. What is important is the algorithm and the quality of the measures obtained.

2.2.6   Stereoscopic Cameras

Stereoscopic cameras are based on photogrammetric principles to obtain the distance from a series of points. Once the common characteristics are detected in two pictures taken from two separate points, the position of the common characteristics can be obtained as well as referencing the whole pictures. What is obtained is a map of distances with the same resolution as the cameras used. This information is generally used for obstacle avoidance, although it can be used also for 3D mapping of the environment.

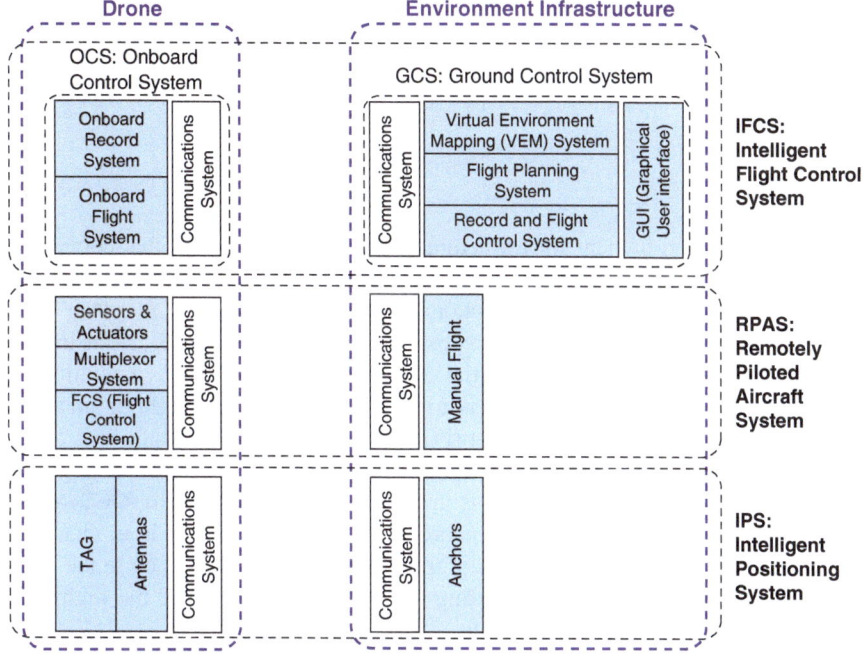

Fig. 2 Indoor Drone distributed system main areas (Drone and Environment Infrastructure) and the three systems involved in each area (IFCS, RPAS and IPS). In each system, the components must connect to others creating an interesting distributed system. Source: own elaboration

3   Indoor RPAS System Architecture

A drone is a distributed system composed of several devices that are connected to provide the whole service of recording video or photographs in indoor scenarios using RPAS. Principally, there are two main working areas: Drone and Environment Infrastructure. Each area has common components in the same three systems: Intelligent Flight Control System (IFCS), RPAS system, and IPS system (see Fig. 2).

3.1   Environment Infrastructure

This is the ground system in charge of supporting the drone operability. It is composed of several subsystems: the IPS anchors that give support to drone indoor positioning, the remote control via radio to control the flight manually, and the Ground Control System (GCS). The GCS is in charge of receiving the whole cloud of points of the scene and surroundings from the Virtual Environment Mapping (VEM) Manager, to generate the flight plan by means of the Flight Planning System.

Finally, the Record and Flight Control System controls and monitors the drone flight.

3.2 Drone

Similar to the Environment Infrastructure, Drone has components that correspond with the previous presented three subsystems. Concerning the IPS, Drone incorporates four antennas to receive the anchors signal, and one board (caller "tag") that processes the signals and generates the position. Related to the RPAS, Drone include all sensors like distance or the Inertial Measurement Unit (IMU), a multiplexor system that provides the source that control the drone (manually flight or automatic flight), and the Flight Control System (FCS) that is in charge of controlling all the parameters of the flight: drone position and orientation, camera parameters, gimbal parameters. Finally, the third system is the On-board Control System (OCS) which also includes the VEM manager (synchronised with the GCS). It is in charge of detecting the cloud of points in front of the drone, sending them back to the GCS, receiving the flight plan and transferring it to the FCS, controlling the flight plan according to the operator requirements.

Concerning the record system, there is a Gimbal, a Record Camera and a VEM Camera, which is usually an RGBD camera (Munera et al. 2015). Gimbal is a specific actuator. It is in charge of orienting the Record Camera to its correct point of interest regardless of the position and orientation of the drone. The Recording Camera (RCam) is in charge of recording the high-resolution video using professional parameters. VEM Camera is a ZCam that provides the drone with the cloud of points of the objects in front of the camera. Both Gimbal and RCam are directly controlled by the FCS using a flight plan or in real time when performing a flight via the OCS that is commanded from the GCS. This paper only focuses on the VEM management both in the OCS and in the GCS during the scanning of the environment phase.

The virtual map of the environment where the drone has to fly is obtained from the same drone used for the recording phase. This carries several advantages: there is only one single drone for all of the recording operation; there is no need to use different devices for different actions (calibration, scanning, tuning and recording); and the size of the equipment to carry, store and move is lower, and consequently, the operation is simplified.

4 Intelligent Flight Indoor Drone Navigation

Safety is the main axis of system design. There are several layers of safety through-out the whole architecture. The lowest level of safety is at the hardware level. It concerns all the devices selected for the hardware, connectors for transmitting the data from one module to another one, power cables, motors, batteries and so on.

There is no control over the basic software layer composed of drivers, O.S. and the low-level FCS. This is not the aim of this work. On this layer, there is another level in charge of the way the drone behaves when it is flying. There are several ways of flying, depending on the degree of autonomy of the drone and the level of safety of the mission to be accomplished:

1. Manual Flight Mode (M.F.M.). The pilot has complete control of the drone. The drone movements have no restrictions and the pilot can instruct it to go anywhere, regardless of any sensor reading or map configuration. The pilot has cancelled the reactive mode (see later) and the human pilot has full control of the drone.
2. Reactive Flight Mode (R.F.M.). This is a defensive flight mode, which is performed by the Flight Control System (FCS), taking into account the reading of the proximity sensors. It is a priority flight mode that is always active unless the human pilot expressly cancels it when flying in Manual Flight Mode. It is active by default for all the other kinds of flight modes: Assisted, Mixed and Smart.
3. Assisted Flight Mode (A.F.M.). The pilot controls the drone, and he or she can take the drone out of the established flight plan. The difference with respect to the Manual Flight Mode (M.F.M.) is that the reactive mode is engaged, so the pilot cannot crash into the environment even if he or she tries.
4. Deliberative Flight Mode (D.F.M.). This is an A.F.M. where the drone is not allowed to move into no-flight zones like populated zones, hanging cables areas and similar spaces.
5. Mixed Flight Mode (Mi.F.M.). The user can explicitly stop and move forward or backward at different speeds along the flight plan. The metaphor is like having a virtual rail along the trajectory of the flight plan. The drone behaves like a 3D virtual dolly. It requires the virtual map to have been captured and the flight plan defined.
6. Guided Flight Mode (G.F.M.). Automatic flight that considers obstacles, restricted areas and surrounding architecture. This mode is completely autono-mous. It is supervised by humans, but humans do not control the drone. A human pilot can pass to any other kind of flight mode from this one. The drone moves automatically along the trajectory of the flight plan. The drone behaves like a 3D virtual dolly as in Mi.F.M., but is completely autonomous. It has the same requirements as Mi.F.M.
7. Smart Flight Mode (S.F.M.). This mode can be engaged when the drone has left the flight plan, and the pilot wants it to return to the predefined flight plan, selecting the shortest itinerary and considering obstacles, restricted areas and surrounding architecture in real time. This mode is completely autonomous. It is supervised by humans, but humans do not control the drone. The implementation

of this mode is beyond the goals of this work. It requires the virtual map, the flight plan and 3D sensors.

8. Emergency Flight Mode (E.F.M.). Moreover, in the event of a loss of IPS datalink, radio contact, engine failures or battery level below a safety level, a defensive failsafe behaviour will be executed. Depending on the type of failure and position of the RPAS, it will start a slow landing or return automatically to the starting point (return to launch—RTL Mode). It requires Environment Scanning.

5 Security vs. Flight Modes

There is a relationship between the flight modes of the drone and the security level required. See Fig. 3. Notice that R.F.M. is not an automatic flight mode that is selectable by the user but rather is a cross-cutting safety feature for all flight modes except for the M.F.M., where the human pilot has full control of the drone. So, it is not included as a flight mode in the horizontal axis.

Notice that the Emergency Flight Mode can only activate the RTL mode without entering in S.F.M. if the drone is in the path. In this mode, the drone can perform an emergency landing in any situation if it does not have enough battery to come back home through the shortest path available. This is an exceptional mode that can be reached from any other state of the drone.

If the drone is not following the flight plan, a return straight line trajectory could be dangerous since the drone could collide with an obstacle or fly into a no-flight zone. In case of detecting an obstacle, the drone cannot decide where to move to. It cannot recalculate in real time an alternative trajectory to RTL. In the case of the battery level being really low, the drone can land on any safe landing point or area specified in the map. This requires an Environment Scanning since this mode requires knowledge of the path and potential obstacles (walls, environment cloud of points, and so on) to determine the return path to a landing point or area (typically the take-off point).

Observe that M.F.M. and A.F.M. may be used in any situation and do not require any kind of scanning of the environment. It is interesting that they are the flight modes currently used when flying a drone in indoor scenarios, using any current on-the-shelf commercial drone. Finally, the drones do not require the use of an accurate IPS, since the flight is completely manual.

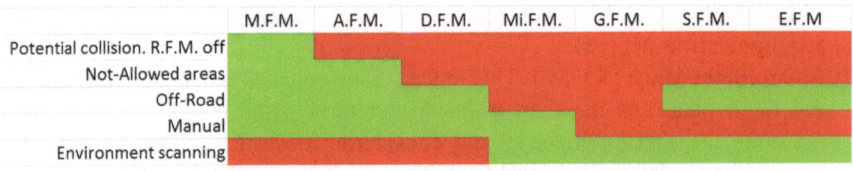

Fig. 3 Relationship between flight modes and security levels. Source: own elaboration

In the case of the D.F.M., this mode does not require any kind of scanning of the surroundings since this flight mode has to avoid restricted areas. The D.F.M. mode is a flight mode that improves security for drone indoor navigation over a completely free M.F.M. since it takes into account both the R.F.M. and the not allowed areas. It is important that indoor spaces cannot use any current on-the-shelf commercial drone since the restriction has to be edited by the pilot in a GUI on a PC/Tablet and later, transferred to the drone FCS in order to avoid those areas when flying. This mode requires use of an accurate IPS since current GPS has a minimum resolution of 10 m when flying outdoors. On the other hand, there are many situations where GPS reception is bad or even non-existent when working indoors.

Notice that (Mi, G, S, E).F.M.:

1. Require scanning of the environment since this flight mode has to avoid not only restricted areas but walls, columns, furniture or any area or volume that the director considers dangerous or not available.
2. Improve security for drone navigation in indoor scenarios, preventing the pilot from crashing into the surrounding walls and furniture.
3. Cannot use any current on-the-shelf commercial drone since forbidden areas, allowed flying paths and flight plans have to be edited by the pilot in a GUI on a PC/Tablet and later transferred to the drone FCS in order to avoid those areas when flying.
4. Require use of an accurate IPS since current GPS has a minimum resolution of 10 m when flying outdoors. On the other hand, there are many situations where GPS reception is bad or non-existent when working indoors.

6   Flight Mode State Machine

A drone may be understood as a state machine. This machine works on different flight modes. Every kind of flight mode may be understood as a state of this state machine. In Fig. 4, there is a state-transition diagram that shows how an indoor drone should work.

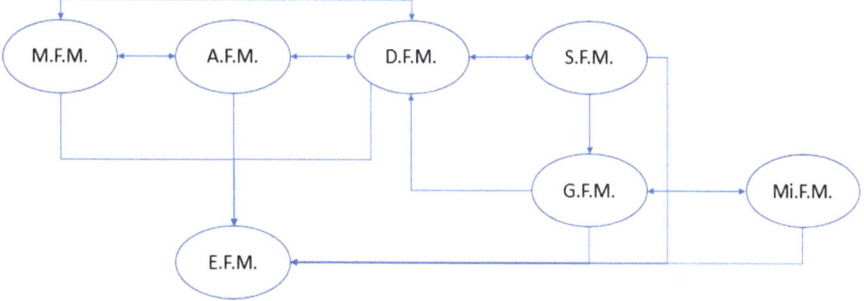

Fig. 4 State-transition diagram of all the flight modes. Source: own elaboration

Notice that there is a group formed by the (M, A, D).F.M. on the left side. This is the manual group, while the group formed by the (Mi, G, S).F.M. on the right side is the automatic one.

The manual group has three characteristics:

1. It is controlled directly by the human pilot. The drone does not move if the human pilot does not order anything (it flies in loiter mode).
2. The human pilot can change from one mode to any other simply by activating the defensive/reactive flight mode or de/activating the no-flight zones. It depends on the safety risks the pilot wants to assume.
3. D.F.M. has the highest security level of all the manual flight modes. It does not allow the pilot to get the drone into no-flight zones and does not allow the drone to collide with walls. So, it is the flight mode reached as soon as the user enters in manual mode while the drone is in G.F.M.

Notice that S.F.M. is a transition state that allows the drone to return back to the original flight plan path in an attempt to avoid obstacles. There are several choices for returning to the original path: going to the nearest path point, to the point where it is supposed to be at that moment according the flight plan, going to the point where the drone left the flight plan, or to a given checkpoint. This mode is the only way to return from a manual flight mode to an automatic flight plan. While the drone is turning back to the original path (S.F.M.), the user can take control of the drone again, passing to D.F.M., the next lower security level. This is why there is a bidirectional arrow between both flight modes. It is the responsibility of the pilot to later reduce the security passing to (M,A).F.M. Once the drone has reached the flight plan trajectory, it enters into the G.F.M. and it starts to follow it as if nothing had happened.

Observe that the automatic group is not directly controlled by the human pilot except for in Mi.F.M., which is a semiautomatic mode where the drone cannot move outside the path in the flight plan. So with this, there is no need to go to S.F.M. from G.F.M. to correct anything. The arrow is one way. Consequently, the human pilot can move the defined trajectory of the flight plan forward and backward by changing from G.F.M. to Mi.F.M. Additionally the drone can move out of the trajectory by changing from G.F.M. to D.F.M.

See that for current implementation, the S.F.M. is not available. So, when moving manually out of the path, the drone is always on D.F.M. and it cannot come back to G.F.M. until the drone is landed and reset.

The diagram shows that manual flight is always mandatory over any other flight mode. Manual flights are always available. They may be seen as an escape mode when the drone is in a risky situation.

Some final remarks about the flight modes: it is important that the transitions from any state to another one are controlled from the Flight Path Manager at the land base. Also, if the manual radio control base is touched, either explicitly or accidentally, the drone will abandon any automatic flight mode and get into D.F.M. for security. Finally, once the drone is in any manual flight mode, the pilot can change to any other manual flight mode by selecting the new mode in the Flight Plan Manager device.

7 Conclusions

This chapter has presented the characteristics of the drones and the specific aspects to be considered in the case of the drone flights in an indoor environment. These characteristics are architecture-oriented in order to achieve the autonomous navigation of indoor drones whose mission is to record videos and pictures with very high definition cameras. In order to achieve this type of mission, the system requires different flight modes that have been proposed: manual, reactive, deliberative and intelligent. Safety is the main axis of system design and a relationship between the flight modes of the drone and the security level required has been presented. Finally, the paper shows how an indoor drone should work by using these flight modes.

References

Castillo P, García P, Lozano R, Albertos P (2007) Modelado y estabilización de un helicóptero con cuatro rotores. Revista Iberoamericana de Automática e Informática Industrial RIAI 4(1):41–57
Hussein A, Al-Kaff A, de la Escalera A, Armingol JM (2015) Autonomous indoor navigation of low-cost quadcopters. In: Alimi AM, Qiu R, Zeng D (eds) Service operations and logistics, and informatics (SOLI), 2015 I.E. international conference. IEEE, Hammamet, pp 133–138
Martínez SE, Tomás-Rodríguez M (2014) Three-dimensional trajectory tracking of a quadrotor through PVA control. Rev Iberoam Autom Inform Ind RIAI 11(1):54–67
Munera E, Poza-Luján JL, Posadas-Yagüe JL, Simó-Ten JE, Noguera JFB (2015) Dynamic reconfiguration of a RGBD sensor based on QoS and QoC requirements in distributed systems. Sensors 15(8):18080–18101
Santana LV, Brandao AS, Sarcinelli-Filho M, Carelli R (2014) A trajectory tracking and 3d positioning controller for the ar. drone quadrotor. In: ICUAS Association (eds) Unmanned Aircraft Systems (ICUAS), 2014 international conference. IEEE, Orlando, pp 756–767

The Relationship of the Industry with the Public Administration: Best Practices on Co-regulation for Training

María de-Miguel-Molina, Virginia Santamarina-Campos, and María-Ángeles Carabal-Montagud

Abstract In the different European countries, co-regulation is used to provide theoretical and practical training to the civilian drone pilots. In this chapter, we present the process required to become an authorised training organisation, and we explain how manufacturers and operators are also involved in this task. The assessment of civil drone pilots is delegated from the public administration to non-governmental organisations and therefore trust is a key factor in this public-private partnership. Therefore, in this chapter we reflect on the characteristics of co-regulation within the civilian drone sector and how the industry is involved in this activity to evaluate the pilots' capacity. As we can observe, if public administration gives precise guidelines to the industry, it contributes to better decision-making, although other stakeholders are frequently left out of the process. Moreover, the approval of the European framework is urgent to give more support to the regulation of the sector and to increase the possibilities of co-regulation.

1 Introduction

New public-private models propose joint decision-making between companies and stakeholders such as the public administration (co-regulation). The European Aviation Safety Agency (EASA), as an Agency of the European Union, promotes the highest common standards of safety and develops common safety rules at the European level. This agency and their national equivalents monitor the activity of producers and operators of civil drones, but depending on the size of the drone, this activity could involve regulation measures or not, in which case other alternatives such as co-regulation can be used. The co-regulation tool, although a soft instrument,

M. de-Miguel-Molina (✉)
Management Department, Universitat Politècnica de València, Valencia, Spain
e-mail: mademi@omp.upv.es

V. Santamarina-Campos · M.-Á. Carabal-Montagud
Conserv. & Restoration of Cult. Heritage Department, Universitat Politècnica de València, Valencia, Spain

© The Author(s) 2018
V. Santamarina-Campos, M. Segarra-Oña (eds.), *Drones and the Creative Industry*,
https://doi.org/10.1007/978-3-319-95261-1_10

is a useful public-private alternative for the manufactures and operators of civil drones.

The regulation of drones (UAVs, Unmanned Aerial Vehicles) and, in particular, RPAs (Remotely Piloted Aircrafts) of less than 150 kg depends on their national regulations. These regulations are mainly focused on safety parameters during the design of civil drones, which are applied to the manufacturers and different services offered by the operators. In the case of the civil drones, the point is to achieve the necessary interaction among stakeholders to produce a consensus of a public policy approach in an area where there is considerable uncertainty (Freeman and Freeland 2014).

The European Union has developed some documents in order to clarify the regulation of civil drones. Current national harmonisation actions undertaken by EASA define riskless open and riskier specific categories (Stöcker et al. 2017). And this is the tendency that the new European Regulation seems to follow by soon establishing different categories (De-Miguel-Molina and Santamarina-Campos 2018).

Moreover, in January of 2018 the Council of the European Union presented the final version of the text agreed upon with the European Parliament on the "Proposal for a Regulation of the European Parliament and of the Council on common rules in the field of civil aviation and establishing a European Union Aviation Safety Agency" (Council of the European Union 2018). This regulation seems that it will be approved around April, and it presumes the necessity of more control over specific drones (probably the riskier ones).

According to its Annex IX, section 4.2, "operators of drones shall be registered where they operate any of the following:

(i) unmanned aircraft which, in the case of impact, can transfer to human kinetic energy above 80 Joules;
(ii) unmanned aircraft, the operation of which presents risks to privacy, protection of personal data, security or the environment;
(iii) unmanned aircraft, the design of which is subject to certification pursuant to Article 46(1)".

That is, taking into account the nature and risk of the activity concerned, the operational characteristics of the unmanned aircraft concerned and the characteristics of area of operation, a certificate may be required for the design, production, maintenance and operation of unmanned aircraft and their engines, propellers, parts, non-installed equipment and equipment to control them remotely as well as for the personnel, including remote pilots, and organisations involved in these activities (article 46).

As we can observe, from a starting point, all drone regulations have one common goal: "minimizing the risks to other airspace users and to both people and property on the ground" (Stöcker et al. 2017). Therefore, national regulations frequently cover the following points:

- Technical requirements (regarding the product).
- Operational limitations (concerning the flight).
- Administrative procedures (certificates, registration, insurance).

- Human resources requirements (qualification of pilots).
- Implementation of ethical constraints (data protection and privacy).

In this case, we will focus on administrative procedures and human resources requirements, but the rest of the points can be found in de-Miguel-Molina and Santamarina-Campos (2018).

To reach a common legal framework, the European Union has developed several stakeholder consultations, although no legislation has been approved yet. Industrial manufacturers and professional users are expected to play a key role and contribute to the decision as to whether drones are going to be a tool for everyone or just for professionals.

In our case of study, the assessment of the civil drone pilots, we find this to be a key factor for the industry (Clarke 2016). Moreover, requiring pilots and operators to be licensed and have insurance can impose standards and ensure safety even when they are not compulsory.

2   The Concept of Co-regulation and the Case of the Training for Civil Drone Pilots

When a non-governmental institution participates with the public administration in regulating a sector, this task can be called "co-regulation". As Höffe (2007) highlights, this is an expansion of the citizens' participation. When some organisations, or even citizens, can apply their power in particular themes, it can be stated that they work in a "transition zone" among the government and the civic (or civil) society. That is, the Third Sector's zone (Catalá Pérez 2017).

The Australian Communications and Media Authority (ACMA 2011:5), points out that "co-regulation generally involves both industry and government (the regulator) developing, administering and enforcing a solution, with arrangements accompanied by a legislative backstop. Co-regulation can mean that an industry or professional body develops the regulatory arrangements, such as a code of practice or rating schemes, in consultation with government. While the industry may administer its own arrangements, the government provides legislative backing to enable the arrangements to be enforced".

Other authors refer to this kind of participation as "meta-regulation" or "meta-governance". According to Peters (2010:37), "it recognises the need for some delegation and devolution of governing but at the same time recognises the need for greater central direction".

Moreover, Sanderson (2011) presents co-regulation as the way of "sharing regulatory responsibilities between the state and regulatees". He emphasises more features of co-regulation:

- It operates within a legislative framework which empowers the regulator to take action in cases of non-compliance,

Table 1 Co-regulation characteristics of the theoretical and practical training for civilian drone pilots

Characteristic	Definition	Application
Target	Entity which the regulation applies	Civil drone pilots, operators, manufacturers, and ATOs[a]
Regulator	Entity that creates and enforces the rule of regulation	Government, Aerial Agencies
Command	What to do or what refrain from doing	Acquired all the documents to prove training
Consequences	What happens if command is not followed	Administrative sanctions for the protection of public safety

Source: own elaboration from Coglianese and Mendelson (2010)
[a]Authorised training organisations

- It enhances legitimacy as regulatees are also involved,
- It relies on high levels of trust between regulator, regulatees and citizens,
- The regulator manages the process as an "equilibrator".

Clarke and Bennett Moses (2014:268) give different power to the actors of a specific regulatory form and, in the case of co-regulation, they stress that the state and the industry "negotiate what and how", while corporations contribute to the industry and other stakeholders "may or may not have some influence". But, from their point of view, in the European drone sector, neither EASA nor the European Commission "show much evidence of engagement with stakeholders outside the industry" (278).

Following Coglianese and Mendelson's (2010) essentials characteristics of regulation, in Table 1 we show how the co-regulation of the theoretical and practical training of civil drone pilots could be considered.

On the other hand, co-regulation can present negative consequences (Peters 2010), which we have analysed:

- Decision-making can be influenced by the interests of the separate parties involved. However, we do not think that this applies to the pilot training, as the requirements of the examinations are quite specific by the regulation. Therefore, decision-making is in fact very limited.
- Some stakeholders are less capable of influence decisions. As highlighted by Clarke and Bennett Moses (2014), we agree in some way. Even if the European Commission has made efforts to open consultations with different stakeholders, National regulations do not always make this process possible.
- More organisations need coordinating. Of course, that could be a problem, but again the regulation can be precise enough to avoid this risk.
- Finally, this mechanism may not substitute for the government responsibility for public action. We are of the same opinion that, in the case of an accident with a drone, public administration is co-responsible if the training approved did not fulfil the standards.

In this sense, we think that some national regulations can be taken as "best practices", although the way to regulate in each country could differ from one to another. Anyway, we should "recognise that greater flexibility will be needed for operating in an environment of innovation and constant change" (ACMA 2011:11).

3   Analysis

Our study is based on a content analysis from different sources of information: academic papers, regulation proposals from the European Union and the regulation of some European countries (mainly, we will focus on the Spanish current regulation, which was the last to come into force in the EU in December 2017). From a comparative analysis of the results, we evaluate the different co-regulations of the National Laws.

As we observe, co-regulation is normally used to provide theoretical and practical training to the pilots of civilian drones. For example, in Spain the National Agency of Aerial Safety (AESA 2017a) works with different organisations to provide theoretical and practical training to pilots. But we think that this tendency could increase in the future, as the industry has many concerns about the role of the public administration.

Therefore, first of all, and following the ACMA assessment framework (2011:13–15), we analyse in Table 2 whether "optimal conditions" or factors are present for co-regulating of the drone sector.

After this analysis, we can conclude that the majority of optimal factors are present, although some of them need to be enhanced. For example, Image 1 shows which risks were perceived by the industry in Spain in 2016, whereas two first risks were "a clear legal framework" and the "slowness of public administrations".

According to ACMA (2011) it is not necessary to cover all of them, but we suggest that if more co-regulation (not only for training) or self-regulation would like to be applied, more focus will be had on the regulatory scheme factors.

It is possible that these two concerns were the starting point to approve the Spanish Royal Decree 1036/2017 BOE, Government Official Bulletin, Royal Decree 1036, 29/12/2017), even before a new European regulation will come into force. This could be, at the moment, the latest national regulation for civil drones in Europe. At any rate, it does not separate the requirements into categories, but still distinguishes drones up to 25 kg from drones that are between 25 and 150 kg.

This new regulation, however, gives the same authorisation for training to:

- Drone manufacturer or organisations authorised by a drone manufacturer.
- Licensed operator with own pilots.
- An authorised training organisation (ATO).

For being considered an ATO, article 5.h of the Royal Decree 2017 includes four categories:

Table 2 ACMA assessment framework applied to the drone sector in Europe

Environmental conditions	Number of market players and coverage of the industry	A small number of players with wide industry coverage will facilitate effective co-regulatory arrangements	The size of drone sector is small compared to others
	A competitive market with few barriers to entry	Co-regulation is less effective where there is little competition or where there is one large player commanding significant market power that cannot be offset by the rest of the industry	The drone sector is quite competitive and many SMEs participate in it. However, they have to be aware of the Chinese competition (DJI as a big competitor)
	Homogeneity of products	Co-regulation is less effective where the products in question are varied and difficult to compare, leading to information asymmetry and product confusion	Products are very homogenous, differences are based on the sector they provide their services
	Common industry interest	Existence of an industry association that is either representative of the whole industry or gives non-members incentives to join	The industry is grouped into different national and/or international associations as well as influenced by specialised authorities
	Incentives for industry to participate and comply	This can include a product marketing value proposition or customer service advantage. Furthermore, the threat of government intervention may provide a sufficient incentive	Training and maintenance are great incentives for the drone manufacturers, which are normally operators as well. Moreover, penalties can be applied in case of non-compliance
	The degree of consumer detriment	In cases of serious risk to public health or safety, direct regulation may be more appropriate; however, intervention must be proportionate to the level of detriment	Safety and security concerns are present; therefore, the participation of other stakeholders is essential
	Whether it is a rapidly changing environment	Regulation that cannot keep pace with developments will be ineffective, and may have unintended and perverse effects, become irrelevant and thus ignored by	The drone sector is developing very fast, its environment changes quickly

(continued)

Table 2 (continued)

		those intended to be regulated, or become an inappropriate mechanism to address its original purpose in a changed environment	
Features of the regulatory scheme	Whether the objectives are clearly defined by the government, legislation or the regulator	It is optimal if policymakers and regulators are clear on what objectives, outcomes and behavioural change they are trying to effect through co-regulatory arrangements. A consistent process for identifying scope, development, enforcement and review is required	There are some differences among countries that are also influenced by the kind of "regulatory culture". Even if the European Commission is putting big efforts for a common framework, the new European regulation has not been approved yet
	Role of the regulator	Does the regulator possess the technical skills to advise on industry proposals? Does the regulator have a clear understanding of the issues? Is data and research available?	In the drone sector, EASA gives the main guidelines and having a specialised agency is a must. This is the same at the National level
	The existence and operation of transparency and accountability mechanisms	The existence and operation of appropriate sanctions to enforce compliance and penalise non-compliance are important indicators of effectiveness. Are scheme members adequately informed about their obligations?	EASA and the National agencies give enough information to the industry, although it is true that this depends on the European country
	Stakeholder participation in the development of the scheme; in particular, consumer input into the development of co-regulatory arrangements	This could be direct participation, such as through consultation processes. Or there could be indirect representation of stakeholder interests, such as through consumer or audience research	The European Commission has undertaken a general consultation to all stakeholders in order to develop the new framework. Anyway, more engagement will be necessary at the National level to apply the arrangements

(continued)

Table 2 (continued)

	Whether the scheme is promoted to consumers	Scheme objectives relating to consumer protection are unlikely to be met if consumers and the community are not made aware of its operation and mechanisms for redress	It will be necessary to give more information, even if insurance is compulsory in many occasions

Source: own elaboration from ACMA (2011)

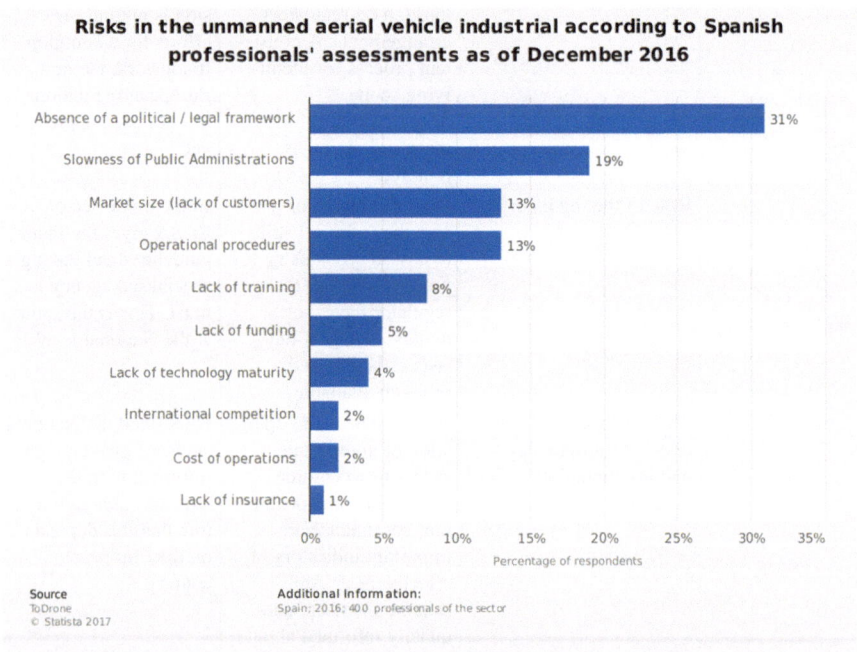

Image 1 Drone-related risks, according to industry in Spain (2016). Source: STATISTA (2017)

- School of ultralight vehicles.
- Non-engine flight school.
- Organisations with AESA qualification.
- Organisations that follow the Regulation (EU) n. 1178/2011 from the Commission, 3rd November 2011. That, at the same time, forward to the Regulation (EC) n. 216/2008.

This latest Regulation (2008) specifies who can assess a civilian drone pilot: on the one hand, experienced operators and, on the other, ATOs. These ATOs must meet the following requirements:

– To have all the means necessary for the exercise of its functions and responsibilities associated with its activity. Among others, these means shall include: facilities, personnel, equipment, tools and material, documentation of tasks, responsibilities and procedures, access to relevant data and registration of data,
– Implement and maintain a safety-related management system and the level of training, and to propose the continuous improvement of this system, and
– If necessary, establish agreements with other relevant organisations to ensure continuous compliance with the above requirements.

After the training and its assessment of the pilot (as described by AESA), these organisations have to send to the Agency a dossier with all the required official documents. In this documentation, the drone type and model that the person is able to pilot should be specified. In Spain, AESA (2017b) publishes its list of ATOs in order to give publicity to the citizens and there we can check the models of drones that they have declared.

The pilot examination comprises three different parts: (a) theoretical knowledge, (b) practical training and (c) medical certificate (AESA 2017a).

Even if this certification is not necessary in all the cases, it could add value in cases of professional works. Moreover, licensed pilots contract insurance and this is another trust guarantee (article 26.c Royal Decree 2017). These licences are covered by many insurance companies, as we can compare in Image 2.

In relation to these three parts for certificating the pilot requirements, Royal Decree 2017 has reduced some requirements in order to diminish regulation but increase co-regulation.

(a) Theoretical knowledge can be demonstrated in two ways (article 34 Royal Decree 2017):

– With a previous licence of any kind for piloting issued by AESA.
– In case of drones up to 25 kg, with a Basic Training Course (for flights within the pilot's visual range) or with an Advanced Course (for flights beyond the pilot's visual range), developed by an ATO according to what is considered the minimum theoretical knowledge that an RPAS pilot should have.

Should an ATO wish to outsource the development of these courses to another organisation, it should include the specific RPAS programmes in its own, monitor and take responsibility for the content and include the trainers in its RPAS-specific instructor cadre. They should also include in their own documentation specific to RPAS the units where the courses are given if they are not their own.

(b) Practical knowledge (article 36 Royal Decree 2017):

A flight book will be the proof that the pilot has enough training. At the end of a practical training course, a successful flight examination can lead to a certificate of

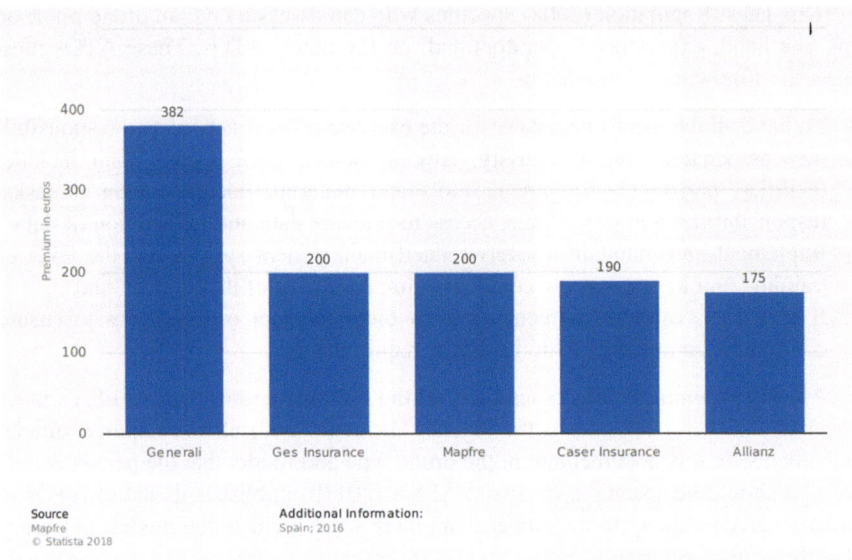

Image 2 Cost of civil liability insurance for drones (UAVs) of less than 25 Kg in Spain in 2016, by Company (in euros). Source: STATISTA (2018)

satisfactory completion, specifying the type and model of aircraft for which the course has been taken. The certificate shall have a footnote indicating the name and position of the issuer. The operator may include in its responsible declaration or modification the aircraft type that can be operated.

This training can be covered by an ATO, an operator or a manufacturer. Pilots should take this flight book with them all the time to demonstrate that in the last 3 months they have performed at least three flights with a specific drone.

(c) Medical certificate (article 35 Royal Decree 2017):

Pilots operating aircraft up to 25 kg maximum take-off mass must hold at least one medical certificate laying down technical requirements and administrative procedures relating to civil aviation flight crew.

Pilots operating aircraft with a maximum take-off mass exceeding 25 kg must hold at least one Class 2 medical certificate, issued by an authorised aviation medical centre or an authorised aerial medical examiner.

Furthermore, in case a pilot would like to be enabled as an operator to give practical training to other pilots, two steps should be covered (AESA 2018a):

Step 1: Presentation of a previous communication to AESA.
Step 2: Carrying out the first flight as an operator to demonstrate that the intended operation(s) with the remote-controlled aircraft can be safely performed.

They shall be conducted and documented in accordance with the AESA guidelines if required by the authority, but shall not be submitted to EASA.

The list of operators is also provided by AESA (2018b) and, at this moment, many of the 2810 operators are also manufacturers.

Other European countries use also co-regulation for training (De-Miguel-Molina and Carabal-Montagud 2018). For example, in the UK, the Civil Aviation Authority (CAA) gives this competence to the National Qualified Entities (NQEs) to conduct commercial operations with drones weighing 7 kg or less. In Belgium, the Direction Générale Transport Aérien (DGTA) gives this competence to some ATOs.

The case is different in Finland, where commercial pilots do not need to apply for any specific licence or certificate, but a report must be made online and manoeuvres documented in a flight book. According to the Finnish Transport Security Agency (TRAFI 2017), they only play an overseeing function, accepting operator notifications, giving permits, handling air space reservations, following occurrence reporting and taking actions on the basis of these information sources. This kind of "supervision" is more in line with the latest Spanish regulation, although some restrictions are still present.

At any rate, the new approach to regulate drones seems to reduce the requirements for training to operators and ATOs, and therefore the ATOs may not be necessary in the future if no licence or certification is compulsory and operator and pilot's responsibility is more focused on personal operation.

4 Conclusions

Even if different regulations, European or National, can distinguish different categories and requirements, we support the idea that operators should have the appropriate training to avoid any risk, even in the cases of small drones. Maybe if the industry is able to develop very precise drones, the pilots could be inexperienced people but, at this moment, we think that these cases should be reduced to indoor environments where the risks can be better assessed.

Although licences and certifications could be reduced in new regulations, a co-regulation where public agencies could give some kind of certificate will be an additional element to reinforce another kind of works where flight licences are not compulsory. Moreover, the necessity of a digital registration of some kind of drones, as proposed by the European Council (2018) or already present in some countries, such as Finland, could be reinforcement for the security of people in case of a drone failure.

The identification and registration of drones seems to be compulsory in the majority of countries, and linking each drone to its owner can help also to assign responsibilities for illegal activities.

As we have observed, by now in the European countries co-regulation is only centred on the operators and practical training. The participation of other stakeholders to ensure safety and security are not included. However, other agencies could be involved with the industry, for example, to ensure information security, product safety or data protection applying different best-practice standards.

In a sector where technology changes frequently, co-regulation will be necessary in order to update the current regulations to reflect the reality. For example, the Spanish regulation has again excluded transport activities even though their inclusion was expected (Sarrión Esteve and Benlloch Domènech 2017).

Moreover, from the side of regulation, the introduction of a compulsory, specific insurance could help. In the same line, citizens see drone regulations as analogous to car regulations, and therefore they should have "mandatory licensing, registration of devices, and mandatory third-party insurance" (Boucher 2016).

We analysed the latest national regulation of drones, the Spanish Royal Decree 2017, as a case study and a best practice to try to give a specific legal framework to the industry in order to avoid ambiguity. We have observed that the Spanish Agency, AESA, is constantly in touch with the industry and updates its information very frequently, which helps with better decision-taking.

References

ACMA (2011) Optimal conditions for effective self- and co-regulatory arrangements. Occasional paper. Australian Government. Available via ACMA. https://www.acma.gov.au/-/media/mediacomms/Research-library-reports-old/Word-Document/Optimal-conditions-for-self--and-co-regulation-Sep-2011-doc.doc?la=en. Accessed 13 Feb 2018

AESA (2017a) Drones, Spanish Agency of Aerial Security, Madrid. https://www.seguridadaerea.gob.es/lang_castellano/cias_empresas/trabajos/rpas/default.aspx. Accessed 7 Feb 2018

AESA (2017b) Authorised training organizations, Madrid. https://www.seguridadaerea.gob.es/media/4357563/listado_atos_rpas.pdf. Accessed 7 Feb 2018

AESA (2018a) Steps to become and operator, Madrid. https://www.seguridadaerea.gob.es/media/4579489/procedimiento_habilitarse_operador_rpas.pdf. Accessed 12 Feb 2018

AESA (2018b) Register of responsible declaration of RPAs operators, Madrid. https://www.seguridadaerea.gob.es/media/4305572/listado_operadores.pdf. Accessed 7 Feb 2018

BOE (2017) Royal Decree 1036/2017, 15 December, of regulation of the use of civil drones. Spanish Ministry of Presidency and Regional Administrations, Madrid. https://www.boe.es/boe/dias/2017/12/29/pdfs/BOE-A-2017-15721.pdf. Accessed 9 Feb 2018

Boucher P (2016) 'You wouldn't have your granny using them': drawing boundaries between acceptable and unacceptable applications of civil drones. Sci Eng Ethics 22:1391–1418

Catalá Pérez D (2017) La colaboración público-privada. In: De Miguel Molina M, Bañón Gomis AJ, Catalá Pérez D (eds) Management para las Administraciones públicas. Editorial Universitat Politècnica de València, Valencia, pp 131–158

Clarke R (2016) Appropriate regulatory responses to the drone epidemic. Comput Law Secur Rev 32(1):152–155

Clarke R, Bennett Moses L (2014) The regulation of civilian drones' impacts on public safety. Comput Law Secur Rev 30(3):263–285

Coglianese C, Mendelson E (2010) Meta-regulation and self-regulation. In: Baldwin R, Cave M, Lodge M (eds) The Oxford handbook of regulation. Oxford University Press, Oxford, pp 146–168

Council of the European Union (2018) Proposal for a Regulation of the European Parliament and of the Council on common rules in the field of civil aviation and establishing a European Union Aviation Safety Agency, and repealing Regulation (EC) No 216/2008 of the European Parliament and of the Council. http://data.consilium.europa.eu/doc/document/ST-5218-2018-INIT/en/pdf. Accessed 13 Feb 2018

De-Miguel-Molina M, Carabal-Montagud MA (2018) Legal and ethical recommendations. In: De-Miguel-Molina M, Santamarina-Campos V (eds) Ethics and civil drones. Proposals for the industry. Springer, Amsterdam, pp 77–86

De-Miguel-Molina M, Santamarina-Campos V (2018) Ethics and civil drones. Proposals for the industry. Springer, Amsterdam

Freeman PK, Freeland RS (2014) Politics and technology: U.S. polices restricting unmanned aerial systems in agriculture. Food Policy 49(1):302–311

Höffe O (2007) Ciudadano económico, ciudadano del Estado, ciudadano del mundo. Katz, Buenos Aires

Peters BG (2010) Meta-governance and public management. In: Osborne SP (ed) The new public governance? Routledge, London, pp 36–51

Sanderson P (2011) The citizen in regulation. A report for The Local Better Regulation Office. Technical report. University of Cambridge. Available via GOV.UK. https://www.gov.uk/government/uploads/system/uploads/attachment_data/file/262583/11-1473-citizen-in-regulation.pdf. Accessed 7 Feb 2018

Sarrión Esteve J, Benlloch Domènech C (2017) Rights and science in the drone era: actual challenges in the civil use of drone technology. Rights Sci:117–133

STATISTA (2017) Risks in the unmanned aerial vehicle industrial according to Spanish professionals' assessments as of December 2016. Statista-The Statistics Portal. Available via STATISTA. https://www.statista.com/statistics/775963/drones-risks-from-the-industry-according-the-professionals-spanish-people/. Accessed 12 Feb 2018

STATISTA (2018) Cost of civil liability insurance for UAVs of less than 25 kg in Spain in 2016, by company (in euros). Statista-The Statistics Portal. Available via STATISTA. https://www.statista.com/statistics/773501/insurance-from-responsibility-civil-for-drones-cousin-by-insurance-carrier-spain/. Accessed 12 Feb 2018

Stöcker C, Bennett R, Nex F et al (2017) Review of the current state of UAV regulations. Remote Sens 9(5):459–485

TRAFI (2017) Unmanned aviation. https://www.trafi.fi/en/aviation/unmanned_aviation/faq. Accessed 12 Feb 2018

Innovative Strategies for the European SMEs: AiRT Project Main Remarks

Virginia Santamarina-Campos and Stephan Kröner

Abstract The aim of this chapter is to summarise the main achievements of AiRT project, that have been presented in this book, and to provide thought-provoking impulses on the bases of our experiences as well to the IT companies as to CIs to "think outside of the box". By doing so, we demonstrated that both the areas of IT innovation and creative companies can benefit mutually. In fact, we are of the opinion that big part of "real" innovation can only be achieved by transferring knowledge between different sectors and adapting the corresponding technology or approaches to the needs of the demanding sectors. This requires inevitable inter-/ or transdisciplinary teams that arise out of given circumstances, which might be composed of just a few experts from different sectors working together on a common goal.

1 Introduction

The use of Design Thinking allowed us to generate a disruptive innovation. Being able to experiment, as a tool that in origin was born as something residual, focused on creative industries, it has been able to transform an innovative product for new industries (Christensen 2011). This methodology has provided a change of mentality in the interdisciplinary team, having a great positive impact on both the consortium and its results and approach to exploitation.

This is how the AiRT project was born, when a real need by a certain part of the creative industry was detected, which could only be solved by an interdisciplinary team with innovative approaches. With this book, written together with the whole consortium involved, we are happy that we could share our problem-solving

V. Santamarina-Campos (✉)
Conserv. & Restoration of Cult. Heritage Department, Universitat Politècnica de València, Valencia, Spain
e-mail: virsanca@upv.es

S. Kröner
Universitat Politècnica de València, Valencia, Spain

V. Santamarina-Campos, M. Segarra-Oña (eds.), *Drones and the Creative Industry*,
https://doi.org/10.1007/978-3-319-95261-1_11

approach and findings with others, be it research centres/universities or companies in general (and SMEs in particular).

2 Conclusions

The creative and cultural industries (CCIs) have shown exceptional resilience to the economic crisis and are well-placed to grow further in the future due to their role as forerunners in digital innovation (EY 2014).

Although the typical companies of the creative industry are usually small businesses (with few employees), their economic impact should not be undervalued. The creative and cultural industries contribute more than 4% of the GDP of the EU, with 535.9 billion euros in turnover and employ more than 7 million people (EY 2014). This value is often underrated by the general public, or rather unknown, and therefore the economic impact on the market is also often underestimated (apart from the cultural value added). Rafa Boix and Pau Rausell illustrated this very comprehensively with the hard data provided in chapter "The economic impact on the creative industry in the European Union".

From the perspective of an economist, the launch of a new product (NPD), can only be successful when there is a real need in the society, or at least if a need can be generated by this new development. The following chapter of the book thus addressed exactly this topic: the need(s) of the creative industries. The authors, Blanca de-Miguel-Molina and Marival Segarra-Oña, very concisely described the needs of the creative industries and how the market for their services is increasing. They clearly identified the targeted markets, the potential of those and the obstacles to overcome. When it comes to the creative industry sector one should be aware of some important facts:

– with over 900,000 companies in Europe the targeting size is big
– it is what economists call an atomised market
– firms are typically composed of about five employees (small sized)

This leads to typical drawbacks concerning innovation processes (lack of resources), but at the same time the SME character of the firms can be converted as an advantage. The company's structures are more flexible, and innovation processes can be applied much faster compared to global players. The AiRT project takes all these considerations into account. We detected a need by the CIs to use creative filming tools such as drones (RPAS) not only outdoors, but also for indoor footage. By doing so, we demonstrated that CI can enhance their services in order to improve their market position. Moreover, the limited resources (available investment amount) are also taken into account, since AiRT will be released at an affordable price, which will be an important characteristic of our value proposition.

By exposing other cases of the use of innovative tools and technology, the possible impact of innovative products on the services offered in the CI field has been demonstrated (see chapter "Succesful cases of the use of innovative tools and technology in the creative industries field"). Technology transfer has had, so far and will have in the future, a strong impact on the film industry as a whole. Digital recording, motion graphics and post-production process can be cited just as a few more examples to illustrate the IT implication which took place in this sector.

It must be highlighted again, that AiRT project's methodology is based on Design Thinking (Both 2009), where the participation of the CIs is present throughout the process. Five main phases were established, whereas the first one detected the needs through the participation of the targeted sectors via focus groups in UK, Belgium and Spain. In the second phase, the information has been synthesised and processed, and subsequently storytelling approach applied, which has been developed in detail in chapter "Storyboarding as a means of requirements elicitation and user interface design: an application to the drone's industry" by Abad et al. It has been shown that this technique is very useful in order to verify whether all the demanded features by CIs can be integrated (or have been integrated already). This was a very important step to assure that the final outcome, the graphical user interface (GUI), meets the required user-friendly interface.

The fourth phase of AiRT methodology can be, from an engineering point of view, described as the most critical/difficult phase of AiRT project: the integration process. Here in this book, first of all, we described in detail the Indoor Positioning System (IPS) in chapter "How a cutting-edge technology can benefit the creative industries: the positioning system at work" (Vermeiren et al.) and the challenges for the integration in a RPAS. The following chapter was been dedicated to presenting the final result after the successful integration process: the indoor drone with advanced features. Why did AiRT project opt for this state of the art (SOA) technique for indoor localisation, based on UWB? As shown in detail in the chapter above, the IPS developed by Pozyx is an affordable solution and is thus accessible for even small sized companies. This has been, since the beginning of the project, an essential requirement. It must be noted and emphasised again, that currently, the most precise IPS solution available is based on motion capture technologies with costs exceeding 200,000 €[1] (VICON n.d.) (just a positioning system), whereas the whole AiRT solution (drone with IPS etc.) is expected not to exceed 10,000 €. Thanks to the implementation of the four-antenna approach and new algorithms, we demonstrated that UWB-based indoor positioning systems are suitable for drone integration and provide sufficient precision to allow professional high-quality filming. Furthermore, again, special emphasis was lain on a user friendly, self-explaining deployment of the system.

What makes our drone different from others? Basically, it can be stated that, up until now, the AiRT solution could be considered as unique when it comes to *professional* indoor filming by creative industries. Since, in the AiRT consortium, the Creative Filmmaking Agency, and thus a targeted end-user, Clearhead is

---

[1]Estimated price. Keep in mind those systems are normally designed for relatively small-sized rooms (6 m × 6 m × 3 m; width/length/height). Larger rooms require more motion capture cameras (prices for each camera are around 2445 €). In typical AiRT use environments (20 m × 20 m × 10 m) more than 100 motion capture cameras might be needed.

involved, the most demanded features for professional footage has been integrated. In addition, the continuous involvement of the creative industries and ICT experts of the MCI ensured success of both need analysis and feasibility of the required features. In particular, the ability to programme flight paths, mark key frames, change (almost) any parameter of the filming camera make up the difference to competitors.

Also, a unique approach—introducing a number of flight modes—provides the AiRT drone with a very sophisticated safety system. The flight modes provide users with different degrees of freedom, from a totally manual flight option to a totally autonomous flight, and a number of intermediate flight modes with configurable restrictions (such as flight at a given altitude or keep a distance to the object in front). This adaptability to different filming environments or drone flying skills by the operator ensures a safe flight at any time.

Although AiRT emphasised the usability of the system in each step, we dedicated the last chapter to the relationship of the industry with the public administration, in particular the co-regulation for training. This is because in the different European countries, co-regulation is used to provide practical training to drone pilots, which is a key factor for the industry. We hope, that we could clarify the process of how to become a drone operator and a pilot and how the industry is involved in this task.

3 Final Remarks

Financing tech-transfer and innovation, as demonstrated in chapter "Financing tech-transfer and innovation: an application to the creative industries" by Cruz et al., especially for small (and medium) sized companies, is an important tool introduced by the European Commission. In particular, Horizon 2020, launched in 2014, is addressed to the topics of research and development. Innovation under such programmes should be tailored to the needs of the end-users (market proximity), proven by demonstration and piloting. While an increase of project funding that involved creative industries was recognisable already in the 7th European Framework Programme (FP7; 2007–2013), an exponential increase of funded projects with CI implication can be observed under H2020.

These funding programmes are particularly important in the area of the creative industry sector, as innovative ideas often exist, but realisation fails due to a lack of resources. This is also true in the case of the AiRT project, where three small SME companies and a university have joined forces to work on a common idea and put it into reality. Without the financial support of the Horizon 2020 programme, the AiRT system would not have been possible. This is why we want to especially acknowledge the funding received from the European Union's Horizon 2020 research and innovation programme under Grant Agreement No 732433.

References

Both T (2009) Bootcamp Bootleg. https://dschool.stanford.edu/resources/the-bootcamp-bootleg

Christensen CM (2011) The innovator's dilemma: the revolutionary book that will change the way you do business. Am Polit Sci Rev 94. Harper Collins, New York

EY (2014) Creating growth: measuring cultural and creative markets in the EU. EYGM Limited, France. http://www.ey.com/Publication/vwLUAssets/Measuring_cultural_and_creative_markets_in_the_EU/$FILE/Creating-Growth.pdf. Accessed 16 Mar 2018

VICON (n.d.) VICON installation. https://www.vicon.com/installation. Accessed 8 Mar 2018

Correction to: Introduction to Drones and Technology Applied to the Creative Industry. AiRT Project: An Overview of the Main Results and Actions

Virginia Santamarina-Campos and Marival Segarra-Oña

Correction to:
Chapter 1 in: V. Santamarina-Campos, M. Segarra-Oña
(eds.), *Drones and the Creative Industry*,
https://doi.org/10.1007/978-3-319-95261-1_1

The original version of first chapter was inadvertently published without including the following funder information: "Program for the promotion of scientific research, technological development and innovation of the Counsel of Education, Research, Culture and Sport, Valencian Region. Reference: AORG/2018/093". The chapter has been updated.

The updated online version of this chapter can be found at
https://doi.org/10.1007/978-3-319-95261-1_1

© The Author(s) 2019   E1
V. Santamarina-Campos, M. Segarra-Oña (eds.), *Drones and the Creative Industry*,
https://doi.org/10.1007/978-3-319-95261-1_12